NOW 2 kNOW!

High Paying Careers in STEM

2nd Edition

by T. G. D'Alberto

Pithy Professor Publishing Company

Brighton, CO

Published by

Pithy Professor Publishing Company, LLC
PO Box 33824
Northglenn, CO 80233

ISBN: 978-0-9882054-7-5

Library of Congress Control Number: 2014955866

Printed in the United States of America

About the Author

Dr. Tiffanie G. D'Alberto has a Ph.D. in Electrical & Computer Engineering from Cornell University and a B.S. and M.S. in Electrical Engineering from Virginia Polytechnic Institute & State University. She has worked for over a decade in the telecommunications and aerospace industries as a scientist, project manager, and supervisor.

In her spare time, Tiffanie enjoys oil painting, drawing, reading, sewing, and running. She's a huge fan of Star Trek, Renaissance Festivals, and animals.

Dedication

To my friend and colleague, Dr. Glenn Bennett, who uses his knowledge and passion for science to mentor and encourage others of all ages.

Acknowledgements

I always thank my family first: My parents for the foundation, the push, and the belief in me all along; My fiancé for his inspiration, encouragement, and unending support.

A huge thanks goes to my friend and colleague, Dr. Glenn Bennett, who sets a great example of how to help others and who has been very supportive in the publishing of this brand.

Finally, I'd like to thank Amazon.com for their excellent publish-on-demand service that enables books such as these, and you, the reader, for making this investment in your future.

Table of Contents

Part 2: Technology   83

Introduction

Welcome!

Choosing a college and a major are huge steps in life that start with an even bigger choice – your future career. Whether you are just graduating from high school or returning to school after working, this decision is one of the most important ones you'll ever make.

So, why consider careers in science, technology, engineering, or mathematics (STEM)? There are several reasons:

1. **STEM careers are in demand and high paying!** STEM careers are getting a lot of attention in the U.S. because companies are scrambling to find qualified candidates. And, because STEM people are in demand, the benefits and pay are really good. In fact, according to the Department of Labor, new graduates with a B.S. in engineering are the highest paid people right out of a four year program.

2. **College is a vocational school.** When you get out of college, you want to find a job that justifies the money you've just spent. STEM majors train you for real jobs that are available as soon as you graduate.

3. **Graduate school is free.** That's right, for many STEM majors, if you decide to progress to a Master's or Ph.D., not only will tuition be covered, but you may also get to teach or do research for the university for added income! It's like getting paid to go to school. I'm not aware of other majors that can boast this feature.

4. **The work is fascinating and impactful.** Whether you cure disease, build the next Mars rover, or invent the new iPhone, your work will be engaging and meaningful.

5. **The career paths are flexible.** With a technical degree, you usually have a variety of options in terms of the technology you work with. Mechanical and electrical engineering are perhaps the most flexible. And, you can work as a technical contributor or a manager. Many companies will even pay you to get your M.B.A.

With that said, why don't more people do STEM?

1. **People are afraid of the math.** First of all, get a tutor, read some books, but never let a fear of math get in your way of a great career. You deserve the best. Second, you may be surprised to learn that many science careers in fact do not require a lot of math.

2. **People believe it's hard.** Well, this is true. The coursework is serious, and it's challenging. But, it's also well worth it. Four years of studying (and, yes, you'll also have time to have fun, too!) is worth a lifetime of good paychecks.

3. **People think there is higher pay in other areas.** Sometimes this is true. But, very long hours, high job dissatisfaction rates, and pure luck often play a large role in those professions. The path to making mega-bucks (i.e., >$500,000/year) in STEM is to get on the ground floor of a start-up or own your own business like Steve Jobs or Bill Gates or many others.

In this second edition, you will find over 80 possible careers with median salaries above $55,000 per year. In addition, Appendix A will give you some guideline questions to think about as you consider each career path. Appendix B gives a helpful discussion on the costs of college, the means to pay for it, and how to make sure it's worth the money.

Whether you choose any of these careers, or find another great fit in another field, I wish you great success!

Terminology & Other Notes:

All wages reported are in U.S. dollars per year.

The median wage is the wage where 50% of employees make less and 50% of employees make more.

The mid-90-percentile wage is the range between the lowest wage earned by the top 90% of employees and the highest wage earned by the top 10% of employees.

All data is taken from the 2014-2015 Occupational Outlook Handbook published by the U.S. Department of Labor and www.myfuture.com produced by the U.S. Department of Defense.

The job of professor applies to all fields and is described in the Mathematics section, Part 4.

This book is written to help you easily and quickly browse the career paths that most interest you. Every career will have the following information readily available:

- **Description**
- **Sample Day-to-Day Tasks**
- **Work Environment**
- **Required Training**
- **Employment Prospects/Pay**

Once you find some careers that peak your interest, I encourage you to go to www.myfuture.com which will provide much more detailed information to help you further explore your options.

And, don't forget to look at the appendices for help choosing a career to align with your interests and guidelines for dealing with the cost of education.

Part 1: Science

A very broad category capturing subjects ranging from biology to chemistry to physics. Math requirements in this field vary greatly. Examples of Science careers include doctor, veterinarian, chemist, environmentalist, and astronomer.

Advanced Practice Registered Nurse

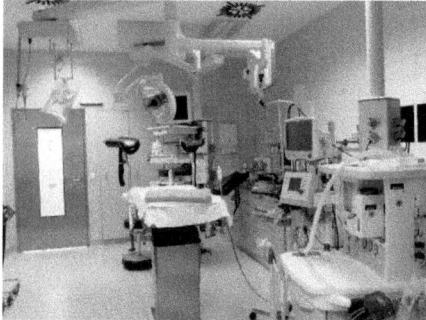

Description:

Advanced practice registered nurses (APRNs) are also called nurse anesthetists, nurse midwives, and nurse practitioners. They can perform the same functions as a registered nurse, but sometimes work independent of physicians. In addition, they also prescribe medication, order tests, and recommend treatments.

Sample Day-to-Day Tasks:

- Examine patients
- Record patient symptoms and histories
- Order diagnostic tests and analyze test results
- Prescribe medications and treatment plans
- Administer medications and treatments
- Counsel patients on conditions and treatment options
- Operate medical and diagnostic equipment
- Consult with physicians as needed
- Assist in research efforts

Work Environment:

APRNs work ~40 hour weeks in hospital, clinic, or home environments. Some positions require shift work or weekend and holiday work. Some APRNs may also be on call depending on the employer and type of specialization. The job can be physically and mentally demanding. Interaction with the public can be rewarding or difficult depending on the client. Work can be found in the military or civilian sectors.

Required Training:

- An M.S. in the area of specialty is the minimum entry level requirement with both classroom and clinical training. Some APRNs also choose to get a Doctor of Nursing Practice (DPN).
- Entry to the M.S. program is only open to licensed registered nurses and typically requires a Bachelor's degree in nursing as well as clinical experience.
- APRNs must also pass a national exam.
- Continual learning through on the job training, journal articles, technical conferences, and classes is expected throughout one's career.

Employment Prospects/Pay:

Job growth is expected to increase 31% over the next decade, about three times the projected average for all occupations, owing to an aging population and increased focus on preventative medicine.

The median salary in 2012:
- $96,460

Mid-90-percentile range:
- $66,000 - $161,000

The average 2012 salary by specialization:
- $148,160 for nurse anesthetists
- $89,960 for nurse practitioners
- $89,600 for nurse midwives

The average 2012 salary by top industries:
- $101,990 in hospitals
- $98,260 in medical practitioner offices
- $97,600 in physicians' offices
- $92,270 in outpatient care centers
- $88,070 in academia

Agricultural & Food Scientist

Description:

Agricultural and food scientists improve the quality and quantity of our food supplies. Specializations include food safety, genetic engineering, water and soil conservation, biofuel development, and increasing a food's appeal and health benefits to the public. Biotechnology for improved yields and nanotechnology for advanced food safety detection are emerging areas of specialization.

Sample Day-to-Day Tasks:

- Design and perform experiments
- Investigate methods of enhanced food production
- Develop engineered plant species for enhanced nutrition and yield
- Analyze soil and water consumption and contamination
- Design methods to analyze food quality and safety
- Use specialized equipment to analyze data
- Communicate results to management and the public in written or oral form
- Publish findings in scientific journals, seminars, and workshops
- Supervise and train personnel

Work Environment:

Agricultural and food scientists typically work 40 hour weeks in office and lab environments. Depending on the position, lab environments may consist of research laboratories, test kitchens, or farms. Work is usually found in the civilian sector, typically in research, government, or commercial food production. Many positions involve research or applied research. Management opportunities are available. Publication & patent opportunities are high.

Required Training:

- A B.S. degree in agricultural science, biology, or chemistry is the minimum requirement to obtain research positions in private industry. Animal scientists typically get a Ph.D.
- An M.S. or Ph.D. is required for university research or faculty positions. Internships are highly recommended for food scientists & technologists.
- Continual learning through on the job training, journal articles, technical conferences, and classes is expected throughout one's career.

Employment Prospects/Pay:

Job growth in this area is expected to increase 9% over the next decade, about the same as the projected average for all occupations.

The median salary in 2012:
- $58,610

Mid-90-percentile range:
- $35,000 - $105,000

The average 2012 salary & top industries for animal scientists:
- $61,580
- Consulting, research & development (R&D), and academia

The average 2012 salary & top industries for soil & plant scientists:
- $58,740
- Federal government, wholesalers, R&D, and academia

The average 2012 salary & top industries for food scientists:
- $58,070
- Management, R&D, food manufacturing, and academia

Starting salaries for new biology baccalaureates in 2009:
- $33,000-35,000

Atmospheric Scientist

Description:

Atmospheric scientists are usually known as meteorologists. Many evaluate current weather conditions and make forecasts based on pressure, humidity, temperature, and wind patterns. Other atmospheric scientists focus on air pollution, global warming, and weather patterns that affect agriculture, defense, and human safety.

Sample Day-to-Day Tasks:

- Use computer aided tools to collect and analyze large amounts of atmospheric data
- Analyze data for trends and forecast impacts
- Perform measurements and maintain databases
- Develop and use mathematical models to make predictions
- Evaluate long range and long term effects of atmospheric phenomena
- Alert public to weather and pollution hazards
- Analyze historic atmospheric data
- Publish findings in scientific journals

Work Environment:

Atmospheric scientists typically work in office and field environments. Depending on the position, field work may include travel. For meteorologists, hours can be long and irregular involving overnights and weekends, again depending on the type of work and the nature of the job. Other types of atmospheric scientists tend to work more normal 40 hour weeks in offices. Work can be found in the military and civilian sectors. Management opportunities are available. Publication & patent opportunities are high in research.

Required Training:

- A B.S. in atmospheric science or related is the minimum requirement for most meteorological positions.
- A Ph.D. is usually required for research or faculty positions.
- Continual learning through on the job training, journal articles, technical conferences, and classes is expected throughout one's career.

Employment Prospects/Pay:

Job growth in this area is expected to increase 10% over the next decade, about the same as the projected average for all occupations. There is likely to be high competition for jobs, and those with higher degrees will fare better.

The median salary in 2012:
- $89,260

Mid-90-percentile range:
- $49,000 - $135,000

The average 2012 salary by top industries:
- $97,710 in federal government
- $86,090 in academia
- $82,360 in radio & television
- $82,310 in science & technological services

Audiologist

Description:

Audiologists work to diagnose and help people suffering from hearing loss or balance concerns. Whether the patient's issue arises from a trauma, a neurological disorder, a virus, or another cause, the audiologist identifies the source of the problem, the severity, and the appropriate treatments.

Sample Day-to-Day Tasks:

- Examine, diagnose, and treat patients
- Use high tech equipment for test and treatment
- Provide hearing screenings
- Work with patients to alleviate symptoms and improve communication with others
- Fit, program, monitor, and repair implants and hearing aids
- Conduct research
- Supervise and train personnel

Work Environment:

Audiologists typically work ~40 hour weeks in office environments. Some weekend and evening work as well as travel may be required depending on the employer. Interaction with the public can be rewarding or difficult depending on the client. Audiologists can work in the military or civilian sectors, in clinics, hospitals, or private practice. Research is a path pursued by some. Publication & patent opportunities are high in research.

Required Training:

- Entry level requires a Master's degree in audiology, though more and more states and businesses are requiring an Au.D. (doctoral of audiology).
- All states require licensing, though requirements vary.
- Continual learning through on the job training, journal articles, technical conferences, and classes is expected throughout one's career.

Employment Prospects/Pay:

With an aging population, job growth in this area is expected to increase 34% over the next decade, more than triple that of the projected average for all occupations.

The median salary in 2012:
- $69,720

Mid-90-percentile range:
- $44,000 - $101,000

Top industries:
- Hospitals
- Physicians' offices
- Clinics
- Schools

Biophysicist or Biochemist

Description:

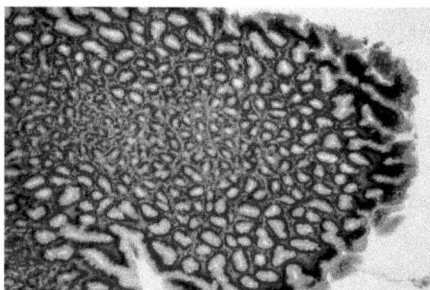

Biophysicists study the impact of electrical and mechanical energy on living cells. They typically work in neuroscience, the study of brain physics, or bioinformatics, the fusion of computer science and biology. Biochemists study the chemistry of living organisms from metabolism to reproduction to growth. Their work is described by the field of biotechnology.

Sample Day-to-Day Tasks:

- Design and perform experiments
- Develop new drugs and medications
- Design and build specialized equipment to analyze data and predict trends
- Create mathematical models and/or computer simulations
- Communicate results to management and the public in written or oral form
- Publish findings in scientific journals
- Supervise and train personnel

Work Environment:

Biochemists and biophysicists typically work 40 hour weeks in office and lab environments. They can work in the military or civilian sectors as well as academia. Most positions involve research or applied research. Management opportunities are available. Publication & patent opportunities are high.

Required Training:

- A B.S. or M.S. degree is the minimum requirement for many applied research jobs or K-12 teaching positions.
- Most researchers are required to have a Ph.D. Post-doctoral work and publications are an asset.
- Coursework involves chemistry, biology, mathematics, physics, engineering, and computer science. These studies aid in running and creating automated tests and computer simulations.
- Continual learning through on the job training, journal articles, technical conferences, and classes is expected throughout one's career.

Employment Prospects/Pay:

Job growth in this area is expected to increase 19% over the next decade, about double that of the projected average for all occupations. There is still likely to be competition for research positions and funding.

The median salary in 2012:
- $81,480

Mid-90-percentile range:
- $41,000 - $147,000

The average 2012 salary by top industries:
- $103,390 in drug wholesalers
- $86,530 in research & development
- $82,490 in pharmaceutical manufacturing
- $74,230 in testing laboratories
- $52,990 in academia

Starting salaries for new biology baccalaureates in 2009:
- $33,254

Chemist or Materials Scientist

Description:

Chemists and materials scientists create new compounds and enhance the knowledge of chemistry. They develop everything from plastics or adhesives to cosmetics or drugs. They can be involved in advancing energy extraction from fossil fuels, creating new circuit chips and superconducting materials, or enhancing food science and pharmaceuticals. If it can be improved upon through chemistry, the chemist is a key player.

Sample Day-to-Day Tasks:

- Plan and execute laboratory experiments
- Research structures and properties of materials
- Confer with customers and management to define requirements
- Develop new materials or compounds to meet customer needs
- Conduct tests to determine long term viability of newly developed products
- Interface with suppliers
- Maintain equipment and instrumentation
- Ensure safe laboratory and manufacturing protocols
- Publish findings in scientific journals

Work Environment:

Chemists and materials scientists typically work 40 hour weeks in office and lab environments. Work can be found in the military or civilian sectors, typically in research, government, or commercial industries. Many positions involve research or applied research, but there are also opportunities in quality control for production facilities. Management opportunities are available. Publication & patent opportunities are high in research.

Required Training:

- A B.S. in chemistry or related is the minimum requirement for most positions.
- An M.S. or Ph.D. is preferred, and a Ph.D. is required for faculty positions.
- Continual learning through on the job training, journal articles, technical conferences, and classes is expected throughout one's career.

Employment Prospects/Pay:

Job growth for chemists and materials scientists is only expected to increase 6% over the next decade with most jobs being offered in the biotechnology sector. There is likely to be high competition for jobs, and those with higher degrees will fare better.

The median salary in 2012:
- $71,770 for chemists
- $88,990 for materials scientists

Mid-90-percentile range:
- $41,000 - $121,000 for chemists
- $47,000 - $134,000 for materials scientists

The average 2012 salary by top industries for chemists:
- $100,920 in federal government
- $79,140 in research & development
- $70,480 in pharmaceutical manufacturing

The average 2012 salary by top industries for materials scientists:
- $106,770 in chemical manufacturing
- $96,630 in research & development
- $96,620 in electronics manufacturing

Starting salaries for new chemistry baccalaureates in 2009:
- $40,000

Chiropractor

Description:

Chiropractors not only help people with back and neck aches, they also treat the whole person with non-drug, non-surgical techniques. Though alignments and pain alleviation are part of their work, they also attack human disease by addressing lowered resistance from neurological responses to spinal joint issues. They also guide patients as to how they can improve their diet or routines to improve their overall health. Chiropractors can specialize if they wish in areas such as sports injuries, orthopedics, nutrition, and diagnostic imaging to a name a few.

Sample Day-to-Day Tasks:

- Examine, diagnose, and treat patients
- Analyze posture and severity of injuries
- Adjust the spinal column and offer additional treatments as needed
- Examine patients' full history and profile to identify underlying causes
- Administer devices for out of office treatments
- Advise patients on preventative and curative care options

Work Environment:

Chiropractors typically work ~40 hour weeks in office environments. The ability to stand for long hours is required, but greater than average stamina or strength is not. Interaction with the public can be rewarding or difficult depending on the client. Chiropractors typically find work in the civilian sector in clinics, hospitals, or private practice.

Required Training:

- Entry level requires a Doctor of Chiropractic (D.C.) and a passing score on licensing examinations.
- After 3 or more years of study in a related field at the undergraduate level, a student may be admitted to a 4 year chiropractic program.
- Licensing varies by state, but all chiropractors must pass all or part of a 4-part national exam.
- Continual learning through on the job training, journal articles, technical conferences, and classes is expected throughout one's career.

Employment Prospects/Pay:

With a growing emphasis on alternative health care, job growth in this area is expected to increase 15% over the next decade, faster than the projected average for all occupations.

Starting chiropractors may earn a small income, but with experience and building a loyal client base, wages can grow rapidly.

The median salary in 2012:
- $66,160

Mid-90-percentile range:
- $31,000 - $143,000

Conservation Scientist or Forester

Description:

Conservation scientists and foresters manage the use and consumption of natural resources such as trees, minerals, and water. They oversee logging, recreational parks, and environmental protection activities. They help keep forests healthy by planning where to plant vegetation and where to harvest trees. They also assist in containing forest fires. Specializations include soil conservation, pest management, and forest economics.

Sample Day-to-Day Tasks:

- Understand and ensure compliance with government regulations
- Develop plans to assist in keeping forests healthy and productive
- Determine planting and harvesting locations
- Negotiate logging and forest leasing contracts
- Direct and assist in natural emergencies such as forest fires
- Develop/coordinate surveys/studies to assess current conditions
- Monitor sites for wildlife population, forest health, and environmental impacts
- Plan conservation projects for wildlife and vegetation
- Communicate results in written and oral form to government agencies and the public as needed
- Supervise and train personnel

Work Environment:

Foresters and conservation scientists work ~40 hour weeks, though the schedule may be non-standard (4 days on, 3 days off, for example). Some work in offices and labs whereas others perform physical outdoor work in the field. Field work can be solitary with long hours, and emergencies will require overtime work. Work is usually found in the civilian sector with about one quarter employed by government agencies. Management opportunities are available.

Required Training:

- A B.S. in forestry, biology, or a related science is the minimum requirement for most positions.
- A Ph.D. is needed for research and faculty positions.
- Continual learning through on the job training, journal articles, technical conferences, and classes is expected throughout one's career.

Employment Prospects/Pay:

Despite the increase in forest fires and demand for wood products, job growth is expected to increase only 3% over the next decade, about one third the projected average for all occupations. Lack of state and federal government funding is likely the culprit for slow job growth in this field.

The median salary in 2012:
- $61,100 for conservation scientists
- $55,950 for foresters

Mid-90-percentile range:
- $38,000 - $91,000 for conservation scientists
- $36,000 - $78,000 for foresters

The average 2012 salary by top industries for conservation scientists:
- $71,110 in federal government
- $53,310 in state government
- $52,820 in social advocacy
- $51,230 in local government

The average 2012 salary by top industries for foresters:
- $64,180 in logging
- $61,680 in federal government
- $56,430 in sawmills & wood preservation
- $54,290 in local government
- $49,610 in state government

Dental Hygienist

Description:

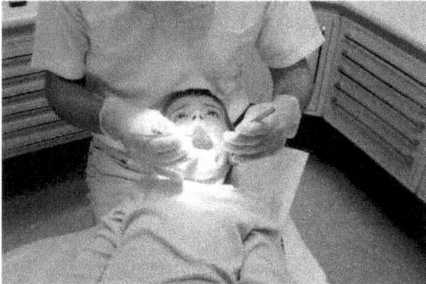

Dental hygienists assist dentists with preliminary examination techniques. They check patients for signs of issues or problems, perform x-rays, and record their findings. They also perform cleaning procedures and instruct patients in preventative care and treatment for existing conditions.

Sample Day-to-Day Tasks:

- Examine patients
- Remove plaque and tartar
- Perform x-rays
- Apply fluorine treatments and coatings
- Provide guidance on preventative and curative measures
- Answer patient questions
- Keep detailed records
- Interact with dentists and dental assistants

Work Environment:

Over half of dental hygienists work part time in dentist offices. Many work for several dentists to increase hours. Interaction with the public can be rewarding or difficult depending on the client. Bending and standing are key aspects of the job. Dental hygienists typically work in the civilian sector.

Required Training:

- Entry level requires an Associate's degree in Dental Hygiene. Some schools may require one year of college before admission to the two year program.
- B.S. and M.S. degrees are also available thought not often pursued unless going into research or teaching.
- All dental hygienists must be licensed. Requirements vary by state and may involve passing a general exam.

Employment Prospects/Pay:

With in increased focus on preventative care, job growth in this area is expected to increase 33% over the next decade, about triple that of the projected average for all occupations.

Full time dental hygienists are eligible for benefits depending on the employer, but over half of dental hygienists work part-time.

The median salary in 2012:
- $70,210

Mid-90-percentile range:
- $47,000 - $96,000

Dentist

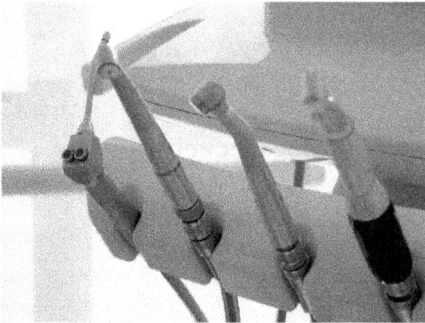

Description:

In addition to striking terror in the hearts of the populace, dentists care for the teeth and oral tissues of their patients. They prescribe preventative care methods, diagnose problem areas, and perform procedures to alleviate pain and future complications. Dentists can be general practitioners or specialize in one of nine areas: Orthodontics (teeth straightening), oral and maxillofacial surgery (surgeons for the head/neck), pediatrics (children), periodontics (gum/bone), prosthodontics (teeth replacement), endodontic (root canal therapy), pathologic (disease), or oral and maxillofacial radiology (x-rays).

Sample Day-to-Day Tasks:

- Examine, diagnose, and treat patients
- Use high tech equipment for test and treatment
- Fill cavities, perform root canals, and extract teeth as needed
- Administer Novocain and write prescriptions
- Provide guidance on preventative and curative measures
- Calm patient fears (and ask them questions when their mouths are full of instruments)
- Supervise and train personnel

Work Environment:

Dentists typically work ~40 hour weeks in office environments. Interaction with the public can be rewarding or difficult depending on the client. Dentists can work in the military or civilian sectors, in clinics, hospitals, or private practice. About 75% of dentists have their own practice.

Required Training:

- Entry level requires a Doctor of Dental Surgery (D.D.S.) or a Doctor of Dental Medicine (D.M.D.) and a passing score on licensing examinations.
- After 2 or more years of study in a related field at the undergraduate level, a student may be admitted to a 4 year dentistry program.
- Competition to get into dental school is high, so 85% of accepted students have completed a baccalaureate degree.
- Licensing varies by state, but all dentists must pass a national exam.
- Continual learning through on the job training, journal articles, technical conferences, and classes is expected throughout one's career.

Employment Prospects/Pay:

With many dentists expected to retire, job growth in this area is expected to increase 16% over the next decade, faster than the projected average for all occupations.

Earnings varied widely by area of specialization, region, and type of practice. Self-employed dentists usually earned more than salaried ones, but a private practitioner has to pay their own benefits and insurance.

The median salary in 2012:
- $149,310 overall
- >$187,200 for oral & maxillofacial surgeons and orthodontists
- $169,130 for prosthodontists
- $145,240 for general dentists

Mid-90-percentile range:
- $74,000 - $187,000

Dietitian or Nutritionist

Description:

Dietitians and nutritionists share their expertise to help people achieve specific goals through their food intake. They design special diets for people fighting afflictions such as obesity, diabetes, or hypertension. Specialties include clinical (working in hospitals or care facilities), management (overseeing kitchens in food service areas), and community (public outreach and education).

Sample Day-to-Day Tasks:

- Educate clients and staff about nutrition
- Assess patients current diet and related health issues
- Design guidelines or meal plans that balance dietary goals, cost, and food preferences
- Track progress of diet prescriptions
- Communicate nutrition issues and solutions to the general public
- Continue personal education in current scientific knowledge
- Supervise and train personnel

Work Environment:

Dietitians and nutritionists work ~40 hour weeks in hospitals, schools, cafeterias, or in offices interacting with individual clients. About 20% work part time. Interaction with the public can be rewarding or difficult depending on the client. Work can be found with an employer or in private practice and usually interacts in the civilian sector. Publication opportunities are available.

Required Training:

- A B.S. in food science is obtained by most dietitians and nutritionists.
- Coursework includes chemistry, biology, and physiology.
- Supervised training in parallel with coursework or in an internship is usually required.
- Some states require licensing.
- Continual learning through on the job training, journal articles, technical conferences, and classes is expected throughout one's career.

Employment Prospects/Pay:

With an ever increasing epidemic of diabesity, job growth is expected to increase 21% over the next decade, about double the projected average for all occupations.

The median salary in 2012:
- $55,240

Mid-90-percentile range:
- $35,000 - $78,000

Top industries:
- Hospitals
- Government
- Care facilities
- Health practitioner offices
- Care centers

Environmental Scientist

Description:

Environmental scientists analyze air, water, and soil samples and develop measures to protect the environment and the surrounding populations. In private industry, they assess the impact of planned projects and ensure new work is in compliance with regulation. In government, they help assess current environmental conditions, reclaim damaged sites, warn people about possible hazards, and help develop the regulations to protect us, the animals, our air, our water, and our land.

Sample Day-to-Day Tasks:

- Collect and analyze water and soil samples
- Perform measurements and determine quality/validity of data
- Make recommendations to ensure minimal environmental impact
- Develop ways to reclaim damaged sites
- Understand and ensure compliance with current regulations
- Communicate results in written and oral form to management, government agencies, and the public as needed
- Maintain equipment and instrumentation
- Investigate/report on events that lead to environmental damage
- Supervise and train personnel

Work Environment:

Environmental scientists work ~40 hour weeks in office and field environments. The amount of field work is very much governed by experience, where junior people may spend a great deal of their time in the field or traveling. Occasional deadlines may require longer hours. Work can be found in the military or civilian sectors with almost half employed by government agencies. Publication & patent opportunities are high. Management opportunities are available.

Required Training:

- A B.S. in an earth science is the minimum requirement for most positions.
- An M.S. is sometimes preferred.
- A Ph.D. is only needed for research and faculty positions.
- Continual learning through on the job training, journal articles, technical conferences, and classes is expected throughout one's career.

Employment Prospects/Pay:

With more emphasis upon understanding environmental issues, job growth is expected to increase 15% over the next decade, faster than the projected average for all occupations.

The median salary in 2012:
- $63,570

Mid-90-percentile range:
- $39,000 - $110,000

The average 2012 salary by top industries:
- $95,460 in federal government
- $67,770 in engineering services
- $64,940 in consulting
- $60,280 in local government
- $56,640 in state government

Starting salaries for new earth science baccalaureates in 2009:
- $39,160

Epidemiologist

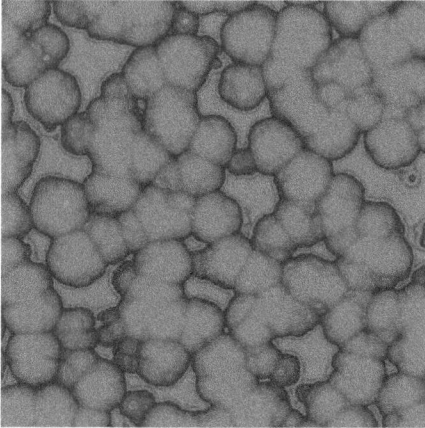

Description:

Epidemiologists determine the cause of diseases or epidemics. Through root cause analysis, they can prevent the spread and recurrence of harmful health events. Epidemiologists not only deal with infectious disease, they also handle issues related to bioterrorism, occupational health, substance abuse, and environmental health.

Sample Day-to-Day Tasks:

- Design and direct studies to track/correct issues in public health
- Collect and analyze specimens such as urine and blood
- Conduct interviews and observations of affected individuals
- Analyze statistical correlations to determine areas most at risk
- Communicate findings to the public, medical community, and government
- Analyze current practices in data collection and analysis and offer methods for improvement
- Continue personal education in field
- Supervise and train personnel

Work Environment:

Epidemiologists work ~40 hour weeks in office, lab, and field environments. Occasional emergencies may require longer hours. Work can be found in the civilian sectors . Over half work in government, with others finding work in hospitals, universities, or research environments. Precautions and safety protocols are enforced to keep employees safe in this profession. Publication & patent opportunities are high. Management opportunities are available.

Required Training:

- An M.S. in epidemiology or closely related field is the minimum requirement for most positions.
- An Ph.D. is sometimes preferred.
- Training includes scientific study in biology, chemistry, and medicine as well as statistics and patient interaction.
- Continual learning through on the job training, journal articles, technical conferences, and classes is expected throughout one's career.

Employment Prospects/Pay:

An ever-growing population and better medical records keeping will contribute to a job growth of 10% over the next decade, the same as the projected average for all occupations.

The median salary in 2012:
- $65,270

Mid-90-percentile range:
- $43,000 - $108,000

The average 2012 salary by top industries:
- $92,070 in research & development
- $73,810 in hospitals
- $66,960 in academia
- $59,090 in state & local government

Genetic Counselor

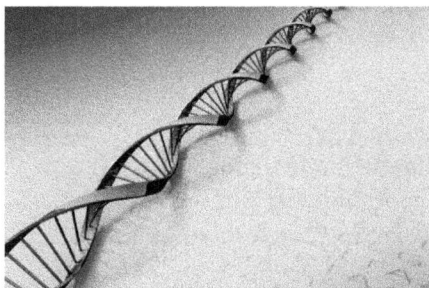

Description:

A genetic counselor offers guidance on the risk of future ailments based on hereditary conditions. Individual or family interviews was well as diagnostic testing are keys to accurate detection of disease. Genetic counselors interface with other physicians to help inform treatment plans.

Sample Day-to-Day Tasks:

- Conduct individual and family interviews
- Answer questions regarding test procedures
- Advise patients on hereditary risks and possible treatments
- Order and analyze lab work and diagnostic tests
- Consult with other physicians to inform treatment plans
- Write reports detailing findings and recommendations
- Keep detailed records

Work Environment:

Genetic counselors work ~40 hour weeks in hospitals, laboratories, physicians' offices, or academic medical centers. They work closely with patients and colleagues. Interaction with the public can be rewarding or difficult depending on the client. Work is usually found in the civilian sector. Publication opportunities are high.

Required Training:

- An M.S. in genetics or genetic counseling is the minimum requirement for most positions. Many candidates earn a Ph.D.
- Certification may be required which entails successful completion of the Master's degree and a passing score on a comprehensive exam.
- Licensing is required in some states . Certification is the first step to licensing.
- Continual learning through on the job training, journal articles, technical conferences, and classes is expected throughout one's career.

Employment Prospects/Pay:

Job growth is expected to increase 41% over the next decade, about quadruple the projected average for all occupations, owing to continued innovations in the field.

The median salary in 2012:
- $56,800

Mid-90-percentile range:
- $26,000 - $88,000

The average 2012 salary by top industries:
- $67,480 in private hospitals
- $63,590 in general hospitals
- $63,240 in academia
- $47,790 in physicians' offices

Geographer

Description:

Geographers study and analyze the Earth in its structure and geopolitical layouts. They update maps and other geographic bodies of knowledge. In addition, they study the peoples and cultures associated with a region and how they are affected by the geography of the area in which they live.

Sample Day-to-Day Tasks:

- Collect/analyze data from direct surveys, satellite imaging, and photography
- Conduct interviews with inhabitants of a region
- Apply statistical methods to their findings
- User computer aided tools to help with their research
- Determine correlations between economics and culture of a people and the physical aspects of a region
- Update geographic bodies of knowledge
- Create visual documentation of results
- Present results to colleagues
- Communicate results in written and oral form to management, government agencies, and the public as needed

Work Environment:

Geographers work ~40 hour weeks in office and field environments with emphasis on the latter. Travel, sometimes to foreign or remote regions, is usually required to conduct interviews and gather data. Work is typically in the civilian sector with most employment found with the federal government. Publication opportunities are high.

Required Training:

- An B.S. in geography is the minimum requirement for entry-level positions in the federal government.
- An M.S. is required for most positions in private industry. Some employers will consider experience an adequate substitute for the advanced degree.
- A Ph.D. is required to teach at universities.
- Study includes statistics and mathematics, and specialization in economics, business, or other sub-disciplines is encouraged.
- Continual learning through on the job training, journal articles, technical conferences, and classes is expected throughout one's career.

Employment Prospects/Pay:

Job growth is expected to be 29% over the next decade, about triple the projected average for all occupations. However, there currently few jobs in this field which means a total of only 500 jobs is expected to be added in this time.

The median salary in 2012:
- $74,760

Mid-90-percentile range:
- $42,000 - $104,000

The average 2012 salary by top industries:
- $78,720 in federal government
- $65,150 in science & technical services
- $53,150 in academia

Geoscientist or Hydrologist

Description:

Geoscientists and hydrologists analyze various physical aspects of the environment, help find natural resources for energy or general consumption, and develop means of protecting our land and water. Geologists study the history of earth formations and piece together the mysteries of evolution (paleontologists fall into this category). Geophysicists study the earth and its components by application of math, physics, and chemistry. Hydrologists study the water cycle and water-related physical phenomena.

Sample Day-to-Day Tasks:

- Collect/analyze data from site surveys, aerial photos, and drilling
- Determine validity/quality of data and interpretations of the data
- Develop predictions of finding gas, oil, and minerals at a given site
- Use computer aided tools to track history and predict trends
- Perform tests using x-ray, heat, acid, and other phenomena
- Develop and coordinate studies to improve knowledge and make informed recommendations
- Communicate results in written and oral form to management, government agencies, and the public as needed
- Maintain equipment and instrumentation
- Supervise and train personnel

Work Environment:

Geoscientists and hydrologists work ~40 hour weeks in office and field environments with emphasis on the latter. Travel is usually required to meet with customers or to conduct field work. Field work can sometimes demand long and irregular hours. Work can be found in the military or civilian sectors with almost half employed by government agencies. Management opportunities are available. Publication & patent opportunities are high.

Required Training:

- An M.S. in an earth science or related is the minimum requirement for most positions.
- A Ph.D. is needed for research and faculty positions.
- Some states require licensing for public work.
- Continual learning through on the job training, journal articles, technical conferences, and classes is expected throughout one's career.

Employment Prospects/Pay:

Job growth is expected to increase 10-16% over the next decade, a little more than the projected average for all occupations.

The median salary in 2012:
- $90,890 for geoscientists
- $75,530 for hydrologists

Mid-90-percentile range:
- $46,000 - $171,000 for geoscientists
- $48,000 - $113,000 for hydrologists

The average 2012 salary by top industries for geoscientists:
- $137,750 in oil & gas extraction
- $94,830 in federal government
- $74,360 in engineering services

The average 2012 salary by top industries for hydrologists:
- $84,540 in federal government
- $80,310 in engineering services
- $78,580 in consulting

Marine Biologist or Limnologist

Description:

Marine biologists study salt water species, and limnologists study fresh water species. Both types of scientists study micro-organisms, animals, and plants. Their work is usually centered at the molecular and cellular level though some scientists in this field also study the physical characteristics of oceans.

Sample Day-to-Day Tasks:

- Design and perform experiments
- Build and maintain databases
- Use specialized equipment to analyze data
- Study organisms at the cellular and molecular level
- Communicate results to management and the public in written or oral form
- Publish findings in scientific journals
- Supervise and train personnel

Work Environment:

Marine biologists and limnologists typically work 40 hour weeks in office and lab environments. Depending on the position, field work may be required which can involve travel, primitive living, and irregular hours. These biologists usually work in the civilian sector, typically in academia. Most positions involve research or applied research. Management opportunities are available. Publication & patent opportunities are high.

Required Training:

- A B.S. or M.S. degree in biology is the minimum requirement for many applied research jobs or K-12 teaching positions.
- Most researchers are required to have a Ph.D. Post-doctoral work and publications are an asset.
- Coursework involves chemistry, biology, mathematics, physics, engineering, and computer science. These studies aid in running and creating automated tests and computer simulations.
- Continual learning through on the job training, journal articles, technical conferences, and classes is expected throughout one's career.

Employment Prospects/Pay:

The following section refers to zoologists and wildlife biologists, which are closely related professions.

Job growth in this area is expected to increase 5% over the next decade, about half the projected average for all occupations. Research positions and funding will be competitive.

The median salary in 2012:
- $57,710

Mid-90-percentile range:
- $37,000 - $95,000

The average 2012 salary by top industries:
- $72,700 in federal government
- $59,670 in research & development
- $57,710 in local government

Starting salaries for new biology baccalaureates in 2009:
- $33,254

Medical Scientist

Description:

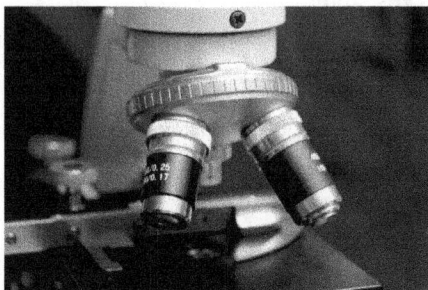

Medical scientists study human disease and develop methods of detection, prevention, and treatment. They develop new drugs, oversee clinical trials, and study the underlying causes of health failures. They may study cells and organisms in the laboratory or work directly with patients extracting blood, administering treatments, or performing invasive procedures.

Sample Day-to-Day Tasks:

- Design and perform experiments
- Develop new drugs and medications
- Assess effectiveness and side effects of treatments and develop ways to improve results
- Take blood or other samples from patients and administer treatments
- Communicate results to management and the public in written or oral form
- Publish findings in scientific journals
- Supervise and train personnel

Work Environment:

Medical scientists typically work 40 hour weeks in office and lab environments. They can work in the military or civilian sectors as well as academia. Most positions involve research or applied research. Management opportunities are available. Publication & patent opportunities are high.

Required Training:

- A Ph.D. is the minimum requirement for most jobs.
- Having an additional M.D. is a great asset.
- Continual learning through on the job training, journal articles, technical conferences, and classes is expected throughout one's career.

Employment Prospects/Pay:

With an aging population and an emphasis on living longer, job growth in this area is expected to increase 13% over the next decade, about the same as that of the projected average for all occupations. There will be competition for research positions and funding in some areas of specialization.

The median salary in 2012:
- $76,980

Mid-90-percentile range:
- $41,000 - $147,000

The average 2012 salary by top industries:
- $92,940 in pharmaceutical manufacturing
- $87,620 in research & development
- $77,180 in physicians' offices
- $71,840 in private hospitals
- $53,740 in academia

Microbiologist

Description:

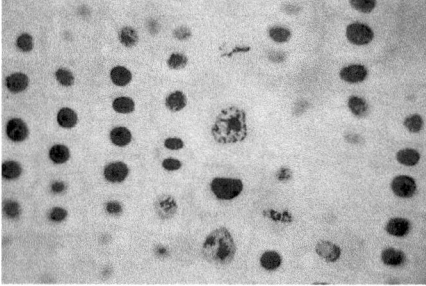

Microbiologists study microscopic organisms such as bacteria and fungi. Specialties include virology, the study of viruses, immunology, the study of disease fighting mechanisms, and bioinformatics, the fusion of biology and computer science for new knowledge generation or new computing methods. Microbiologists impact the environmental, food, and industrial sectors.

Sample Day-to-Day Tasks:

- Design and perform experiments
- Analyze effects of disease and antibiotics on microorganisms
- Isolate and study biological species
- Develop new methods of sterilization for food and pharmaceutical industries
- Use specialized equipment for data analysis
- Perform services for health departments and community programs
- Create mathematical models and/or computer simulations
- Communicate results to management and the public in written or oral form
- Publish findings in scientific journals
- Supervise and train personnel

Work Environment:

Microbiologists typically work 40 hour weeks in office and lab environments. They can work in the military or civilian sectors as well as academia. Most positions involve research or applied research. Publication & patent opportunities are high. Management positions are available.

Required Training:

- A B.S. or M.S. degree in microbiology or a related science is the minimum requirement for many applied research jobs or K-12 teaching positions.
- Most researchers are required to have a Ph.D. Post-doctoral work and publications are an asset.
- Coursework involves chemistry, biology, mathematics, physics, engineering, and computer science. These studies aid in running and creating automated tests and computer simulations.
- Continual learning through on the job training, journal articles, technical conferences, and classes is expected throughout one's career.

Employment Prospects/Pay:

Job growth in this area is expected to increase 7% over the next decade, slower than that of the projected average for all occupations. There is likely to be competition for research positions and funding.

The median salary in 2012:
- $66,260

Mid-90-percentile range:
- $40,000 - $118,000

The average 2012 salary by top industries:
- $96,520 in federal government
- $67,070 in pharmaceutical manufacturing
- $62,920 in research & development
- $54,640 in state & local government
- $52,790 in academia

Starting salaries for new biology baccalaureates in 2009:
- $33,254

Natural Sciences or Medical Sciences Manager

Description:

Natural or medical/health science managers oversee teams of scientists. They help come up with programs to meet executive goals, make sure the projects are adequately staffed, and ensure that employee concerns are addressed. They may take on a more project oriented role in directing activities and taking responsibility for technical, fiscal, and schedule successes. Or, they may operate from a functional standpoint of bridging the gap between staff and executives and having regular career related discussions with employees. In order to be effective, science managers must have a strong background and solid experience in the field they oversee.

Sample Day-to-Day Tasks:

- Develop programs to meet broad goals set by exec. management
- Act as communications bridge between executives and staff
- Develop detailed project plans to meet technical, fiscal, and schedule goals
- Ensure staffing is sufficient on programs and make sure all employees are being utilized equally
- Make presentations for internal and external consumption
- Conduct performance reviews
- Give and solicit regular feedback to and from staff
- Recruit, hire, and fire as required
- Enforce company regulations and set an example of compliance
- Approve time cards, prepare reports, and conduct other administrative duties

Work Environment:

Science managers typically work in offices with ~40 hour weeks, though overtime work can be frequent. Some managers are required to travel to perform on-site inspections or to confer with customers. Jobs can be found in both military and civilian venues. Science managers can work in the research, development, or manufacturing areas. Some also continue to do technical work while others find that their work is limited to management responsibilities

Required Training:

- Entry level requires a B.S. in a related science discipline along with years of experience as a technical contributor.
- Medical and health science managers usually have an advanced degree. Many natural science managers also have a Masters of Business Administration (M.B.A).
- The education and experience requirements will vary by company, but a strong technical background along with an ability to manage people and translate technical jargon to management are required demonstrated skills.
- Continual learning through on the job training, journal articles, technical conferences, and company funded classes is expected throughout one's career.

Employment Prospects/Pay:

Job growth is expected to be 6% over the next decade for natural science managers, and 23% for medical and health services managers.

Pharmaceutical and science research management positions paid the most. The wages below do not include stock options and cash bonuses.

The median salary in 2012:
- $115,730 for natural science managers
- $88,580 for medical and health services managers

Mid-90-percentile range:
- $65,000 - >$187,000 for natural science managers
- $54,000 - $151,000 for medical and health services managers

Nuclear Medicine Technologist

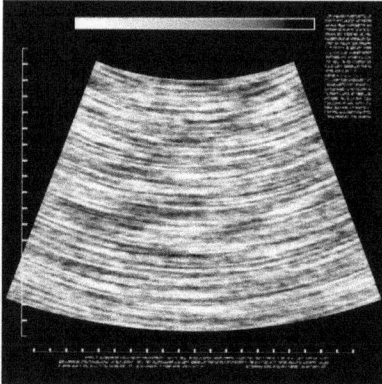

Description:

A nuclear medicine technologist uses scanning equipment to locate areas of abnormal tissue or disease. The technologist administers radioactive drugs to the patient which helps the imaging device detect areas of concern. Only 2-4 years of study is required for this high-paying occupation.

Sample Day-to-Day Tasks:

- Counsel patients regarding questions about test procedures and what can be expected
- Prepare and administer radioactive drugs
- Conduct imaging scans as per physicians' orders
- Operate and maintain imaging equipment
- Interface with computer systems to ensure accurate operation
- Follow safety protocols to protect themselves, staff, and patients
- Monitor patients for reactions to treatment
- Document patient treatment and progress
- Assist is test result analysis

Work Environment:

Nuclear medicine technologists work ~40 hour weeks in hospital and clinic environments. Shift work or weekend and holiday work may be required to handle emergency situations. On-call duties may be required, as well. They typically stand for long hours and need to be physically fit enough to help disabled patients. Interaction with the public can be rewarding or difficult depending on the client. Work is mainly found in the civilian sectors.

Required Training:

- An Associate's degree in nuclear medicine technology is the most common path for entry level work. A B.S. program is also available.
- Some candidates pursue an Associate's or Bachelor's degree in nursing, and then complete a 12 month certification course.
- Some states require licensure with passage of a certification exam.

Employment Prospects/Pay:

With an aging population and increased access to healthcare, job growth is expected to increase 20% over the next decade, about double the projected average for all occupations.

The median salary in 2012:
- $70,180

Mid-90-percentile range:
- $51,000 - $93,000

Top industries:
- Hospitals
- Physicians' offices
- Medical laboratories
- Care centers

Occupational Health & Safety Specialist

Description:

Occupational health and safety specialists analyze work places to assess the likelihood of injury and provide guidance on avoiding hazards. They design programs that ensure the safety of employees as well as the environment by analyzing physical, chemical, and biological threats.

Sample Day-to-Day Tasks:

- Examine work sites to assess hazards
- Perform physical, chemical, and biological analyses on possible threats or contaminants
- Design safety protocols to prevent or minimize threats
- Inspect work sites to ensure compliance with safety protocols and legal restrictions
- Investigate accidents to determine root cause and devise ways to prevent future occurrences
- Supervise and train personnel

Work Environment:

Occupational health and safety specialists typically work ~40 hour weeks in office, lab, or field environments. Travel is usually a key component of the position, and some work involves stressful or dangerous situations. Work can be found in the civilian sector with about a third working in government and the rest working in consulting agencies, hospitals, or manufacturing. Management opportunities are available.

Required Training:

- Entry level requires a B.S. in health and safety, engineering, or a natural science.
- Some positions require an M.S. in health physics or related.
- Coursework involves science as well as studies of laws and inspection procedures.
- Experience is emphasized, so internships are encouraged.
- Certification is not required, but some employers encourage it.
- Continual learning through on the job training or certification classes is expected throughout one's career.

Employment Prospects/Pay:

Job growth in this area is expected to increase 7% over the next decade, a little less than the projected average for all occupations.

The median salary in 2012:
- $66,790

Mid-90-percentile range:
- $40,000 - $97,000

Top industries:
- Federal government
- State & local government
- Consulting
- Education services
- Hospitals
- Manufacturing

Occupational Therapist

Description:

Occupational therapists help people with mental, physical, or emotional disabilities cope with their everyday work and home lives. They enable patients to restore normalcy and self-sufficiency. Whether they support reasoning, motor skills, or the use of assistive equipment, they make a big difference in how people with disabilities relate to the world.

Sample Day-to-Day Tasks:

- Examine, diagnose, and treat patients
- Analyze severity of disabilities and impact on patient life
- Assist patients in learning new skills or use of assistive equipment
- Keep accurate records of hours worked and patient progress
- Use and maintain high tech equipment
- Advise patients on preventative and curative care options
- Supervise and train personnel

Work Environment:

Occupational therapists typically work ~40 hour weeks in patient centers and offices. It is not uncommon to work for more than one employer at a time. The work can be tiring depending on how much standing and lifting of patients is required. Many therapists become supervisors and have staff to perform the hands on work with patients. Interaction with the public can be rewarding or difficult depending on the client. Occupational therapists can work in the military or civilian sectors, in clinics, hospitals, homes, schools, or industry.

Required Training:

- Entry level requires an M.S. or higher in occupational therapy which includes 24 weeks of supervised field work.
- Undergraduate degrees should be in biology or a related field with coursework in other natural sciences, art, and social sciences.
- Licensing varies by state, but all occupational therapists must pass a national exam.
- Continual learning through on the job training, journal articles, technical conferences, and classes is expected throughout one's career.

Employment Prospects/Pay:

With an aging population, job growth in this area is expected to increase 29% over the next decade, about triple the projected average for all occupations.

The median salary in 2012:
- $75,400

Mid-90-percentile range:
- $51,000 - $107,000

The average 2012 salary by top industries:
- $83,430 in nursing facilities
- $82,310 in home health care
- $77,430 in health practitioner offices
- $75,140 in hospitals
- $66,610 in K-12 school environments

Optometrist

Description:

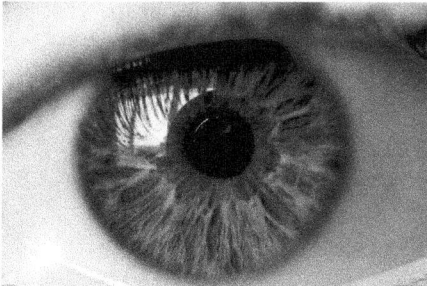

Optometrists are eye doctors. They quantify the degree to which a patient is suffering vision problems such as with depth perception, color vision, near and long range clarity, and proper eye function. They prescribe medications or corrective lenses as needed. They also advise patients on how they can improve their diet or routines to improve their overall eye health.

Sample Day-to-Day Tasks:

- Examine, diagnose, and treat patients
- Analyze clarity of vision, color perception, and eye coordination
- Diagnose other illnesses such as glaucoma as detected by ocular tests
- Use and maintain high tech equipment
- Prescribe medications
- Prescribe corrective lenses and test for fit and comfort
- Advise patients on preventative and curative care options

Work Environment:

Optometrists typically work ~40 hour weeks in office environments. Interaction with the public can be rewarding or difficult depending on the client. Optometrists can work in the military or civilian sectors, in clinics, hospitals, or private practice.

Required Training:

- Entry level requires a Doctor of Optometry (O.D.) and a passing score on licensing examinations.
- After 3 or more years of study in a related field at the undergraduate level, a student may be admitted to a 4 year optometry program after passing the Optometry Admissions Test.
- Competition for optometry school is high with only 1/3 of applicants accepted in 2007.
- Licensing varies by state, but all optometrists must pass a national exam.
- Continual learning through on the job training, journal articles, technical conferences, and classes is expected throughout one's career.

Employment Prospects/Pay:

With an aging population, job growth in this area is expected to increase 24% over the next decade, more than double that of the projected average for all occupations.

Salaried optometrist typically earn more than ones in private practice early in their career, but in the end, self-employment pays off well with about a $20,000 increase in average salary.

The median salary in 2012:
- $97,820

Mid-90-percentile range:
- $53,000 - $185,000

Orthotist or Prosthetist

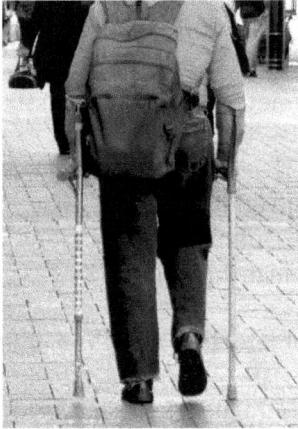

Description:

Orthotists and prosthetists are also called O&P professionals. They design medical aids and instruments and fit patients to them. Orthotists specialize in supportive devices such as braces and inserts. Prosthetists specialize in artificial limbs and body parts. An O&P professional may work in one or both fields.

Sample Day-to-Day Tasks:

- Interview patients to assess their needs
- Measure patients and cast molds to fit them for devices
- Design equipment from a physician's prescription
- Provide adjustments to equipment
- Give instruction to patients
- Repair devices
- Record patient progress

Work Environment:

O&P professionals typically work ~40 hour weeks in office environments. Interaction with the public can be rewarding or difficult depending on the client. Work can be found in the civilian sector, with government, medical offices, hospitals, and medical device manufacturers. Publication & patent opportunities are high.

Required Training:

- Entry level requires an M.S. in orthotics and prosthetics.
- Admission to the M.S. program requires a number of science and mathematics courses in the undergraduate major.
- A one year residency is required with at least 500 hours of training split evenly between orthotics and prosthetics.
- Certification is required and involves passing a national exam for either orthotics or prosthetics.
- Sitting for the certification exam is only allowed after the M.S. and one year of residency is complete.
- If pursuing both orthotics and prosthetics certification, one year of residency must be completed for each.
- Continual learning through on the job training, journal articles, technical conferences, and classes is expected throughout one's career.

Employment Prospects/Pay:

With an aging population, limb loss associated with diabetes and heart disease will contribute to an expected job growth of 36% over the next decade, almost quadruple that of the projected average for all occupations.

The median salary in 2012:
- $62,670

Mid-90-percentile range:
- $34,000 - $111,000

The average 2012 salary by top industries:
- $68,680 in medical manufacturing
- $67,750 in health/care stores
- $66,960 in federal government
- $53,930 in physicians' offices
- $51,950 in hospitals

Pharmacist

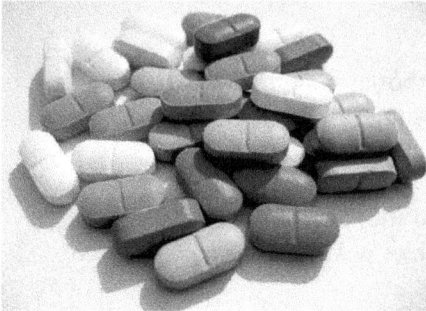

Description:

Pharmacists administer prescriptions, check drug interactions, and counsel patients in correct use and monitoring of their medications. Pharmacists typically work for private industry or for hospitals. Areas of study include chemistry, biology, physics, and math.

Sample Day-to-Day Tasks:

- Review prescriptions
- Check drug interactions and side effects
- Dispense medication to clients
- Compound medical prescriptions for final medication
- Keep careful records
- Advise and counsel patients
- Train patients on at-home medical instrumentation
- Conduct laboratory tests
- Perform administrative duties
- Supervise and train personnel and students

Work Environment:

Pharmacists typically work in comfortable environments, but a lot of standing is usually required. Weekend and night work is not uncommon with "on-call" duties sometimes required, especially in hospitals. Interaction with the public can be rewarding or difficult depending on the client. Pharmacists can work in the military or civilian sectors, usually for pharmacies or hospitals. Management opportunities are available.

Required Training:

- Entry level requires a Doctor of Pharmacy (Pharm.D.) and a passing score on several examinations.
- After 2 or more years of study in a related field at the undergraduate level, a student may be admitted to a 4 year pharmacy program.
- Some students additionally opt for 1-2 years of internship or residency.
- Licensing varies by state, but all pharmacists must pass a national exam.
- Continual learning through on the job training, journal articles, technical conferences, and classes is expected throughout one's career.

Employment Prospects/Pay:

Job growth in this area is expected to increase 14% over the next decade, slightly faster than that expected for all occupations. Training programs currently have limited capacity, so those who make it should have excellent placement opportunities.

The median salary in 2012:
- $116,670

Mid-90-percentile range:
- $89,000 - $146,000

The average 2012 salary by top industries:
- $128,910 in general merchandise stores
- $120,540 in department stores
- $117,850 in pharmacies & drugstores
- $116,000 in grocery stores
- $114,100 in hospitals

Physical Therapist

Description:

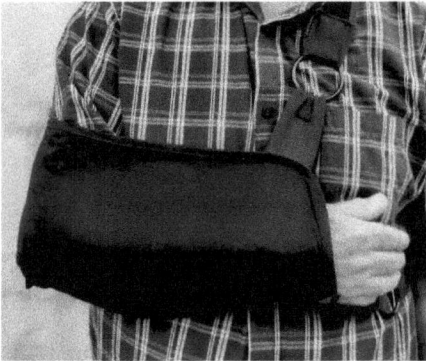

Physical therapists help patients cope with physical disabilities or injuries. They work with people of all ages to overcome anything from sprains or arthritis to burns or stroke. Physical therapists work directly with their patients through hands-on sessions as well as indirectly by prescribing exercises and assistive equipment patients can use on their own. Their focus is on overcoming injury or disability and preventing further loss of function.

Sample Day-to-Day Tasks:

- Examine, diagnose, and treat patients
- Analyze severity of disabilities and impact on patient life
- Assist patients in learning new skills or use of assistive equipment
- Keep accurate records of hours worked and patient progress
- Use and maintain high tech equipment
- Advise patients on preventative and curative care options
- Supervise and train personnel

Work Environment:

Physical therapists typically work ~40 hour weeks in patient centers and offices. The work is usually physically demanding with lots of standing, stooping, lifting, and supporting. Interaction with the public can be rewarding or difficult depending on the client. Physical therapists can work in the military or civilian sectors, in clinics, hospitals, homes, or schools. Management opportunities are available.

Required Training:

- Entry level requires an M.S. or higher in physical therapy which usually includes supervised field work.
- Coursework includes the natural sciences with emphasis on physiology, neuroscience, pharmacology, and pathology.
- Licensing varies by state, but all physical therapists must pass a national exam.
- Continual learning through on the job training, journal articles, technical conferences, and classes is expected throughout one's career.

Employment Prospects/Pay:

With an aging population, job growth in this area is expected to increase 36% over the next decade, almost quadruple that of the projected average for all occupations.

The median salary in 2012:
- $79,860

Mid-90-percentile range:
- $55,000 - $111,000

Top industries:
- Health practitioner offices
- Hospitals
- Home care services
- Nursing & care facilities
- Physician's offices

Physician Assistant

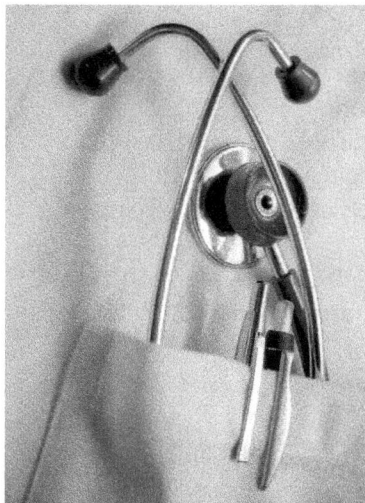

Description:

Physician assistants practice medicine under a doctor's supervision. They are not medical assistants which are professionals who perform administrative and clerical duties in the medical field. Physician assistants test, diagnose, and treat patients. They also prescribe some medications and perform certain medical procedures. Physician assistants are usually found in rural, inner city, or nursing home environments where on-site physicians have limited availability.

Sample Day-to-Day Tasks:

- Examine, diagnose, and treat patients
- Interpret x-rays and lab tests
- Suture, splint, or cast minor injuries
- Assist in surgeries
- Keep accurate patient records
- Order medical supplies
- Supervise and train personnel

Work Environment:

Physician assistants typically work ~40 hour weeks in patient centers and hospitals. Hours can vary and may require weekends, overnights, and on-call duties. Standing and walking are commonly required of the job. Interaction with the public can be rewarding or difficult depending on the client. Physician assistants can work in the military or civilian sectors, in clinics, hospitals, homes, or other care facilities. Management opportunities are available. Publication opportunities are high.

Required Training:

- Entry level requires completion of a 2 year physician assistants training program.
- Admission to the program usually requires a 4 year college degree in biology or a related field along with health-services related experience.
- Many physician assistants are former nurses, EMT's, or paramedics.
- All physician assistants must be licensed and pass a national exam.
- Continual learning through on the job training, journal articles, technical conferences, and classes is expected throughout one's career.

Employment Prospects/Pay:

With a growing effort to reduce medical treatment costs, job growth in this area is expected to increase 38% over the next decade, about quadruple that of the projected average for all occupations.

The median salary in 2012:
- $90,930

Mid-90-percentile range:
- $62,000 - $125,000

The average 2012 salary by top industries:
- $93,660 in hospitals
- $93,520 in care centers
- $90,150 in health practitioner offices
- $88,890 in education services
- $86,870 in government

Physician or Surgeon

Description:

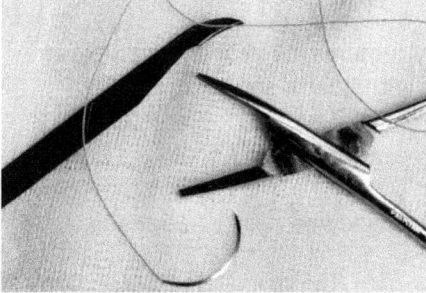

Physicians and surgeons provide medical care to people. They typically work in private practices, clinics, or hospitals. There are many specialists in this field including general practitioners, general internists, anesthesiologists, pediatricians, gynecologists, dermatologists, oncologists, neurologists, etc. Education and training runs the gamut of chemistry, biology, genetics, anatomy, physics, and math.

Sample Day-to-Day Tasks:

- Examine, diagnose, and treat patients
- Prescribe medication, set bones, give injections as needed
- Perform surgery
- Advise and console sick clients
- Operate diagnostic equipment
- Conduct research
- Supervise and train personnel

Work Environment:

Medical doctors typically work in office environments. Hours can be long and irregular. Interaction with the public can be rewarding or difficult depending on the client. Doctors can work in the military or civilian sectors, in clinics, hospitals, or private practice. Research is a path pursued by some. Publication opportunities are high. Management opportunities are available.

Required Training:

- Entry level requires a Doctor of Medicine (M.D.), 3-8 years of internship or residency, and a state license.
- Students typically complete 4 years of undergraduate school in a biology, chemistry, or related field with 4 years of medical school afterward.
- Admission to medical schools is very competitive.
- Licensing varies by state, but all doctors must pass a national exam.
- Continual learning through on the job training, journal articles, technical conferences, and classes is expected throughout one's career.

Employment Prospects/Pay:

Job growth in this area is expected to increase 18% over the next decade, almost double that of the projected average for all occupations. Opportunities are especially high in rural and low-income areas.

The median salary varies significantly among specialties.

The median salary in 2012:
- $220,942 for a primary care physician
- $396,233 for a specialty practice

Other related median salaries in 2012:
- $431,977 for anesthesiology
- $367,885 for general surgery
- $381,737 for ob-gyn
- $224,110 for internal medicine
- $220,252 for psychiatry
- $216,069 for pediatrics and related
- $207,117 for general family practice

Physicist or Astronomer

Description:

Physicists endeavor to understand our universe, from subatomic interactions to the vastness of the cosmos. In addition to the basic sciences, their expertise spans quantum mechanics to relativity with an extensive depth in mathematics. Physicists are researchers that add to our scientific knowledge base. As funding for pure research can at times be tight, physicists also contribute in applied physics areas alongside other scientists and engineers. Work can involve astronomy, particle physics, lasers and optics, and many other fields.

Sample Day-to-Day Tasks:

- Perform complex calculations
- Use computer-assisted analysis tools
- Develop computer simulations
- Design and conduct experiments
- Analyze data
- Publish results to peer reviewed journals
- Collaborate with other scientists
- Teach students

Work Environment:

Physicists typically work in offices and labs with ~40 hour weeks. Jobs can be found in both military and civilian venues. Physicists work in government, private industry, and universities. Work usually centers on research, but applied physics jobs can be found in development areas, as well. Physicists can also perform as technical contributors, program managers, functional managers, and professors. Publication & patent opportunities are high.

Required Training:

- Entry level requires a Ph.D. in physics or a closely related discipline. Post-doctoral work is an asset for jobs in academia.
- Continual learning through on the job training, journal articles, technical conferences, and classes is expected throughout one's career.

Employment Prospects/Pay:

Though competition for pure research funding may be competitive, physicists enjoy a number of other opportunities in the applied science and engineering fields. Job growth in this area is expected to increase 10% over the next decade.

The median salary in 2012:
- $106,840 for physicists
- $96,460 for astronomers

Mid-90-percentile range:
- $58,000 - $177,000 for physicists
- $51,000 - $165,000 for astronomers

The average 2012 salary by top industries for physicists:
- $152,280 in hospitals
- $130,980 in consulting
- $111,020 in federal government

The average 2012 salary by top industries for astronomers:
- $139,140 in federal government
- $93,870 in research & development
- $77,870 in academia

Podiatrist

Description:

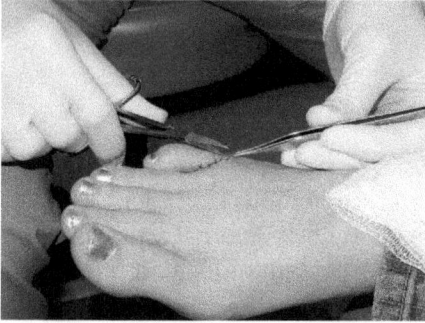

Podiatrists are foot doctors. They treat ailments such as corns, bunions, and ingrown toenails. They repair broken bones and treat deformities and infections. Podiatrists also help people with diabetes who suffer the associated foot issues with that disease. They prescribe medications or corrective orthotics as needed. They also advise patients on how they can improve their diet or routines to improve their overall foot health. In short, they alleviate pain, cure problems, and treat issues associated with a population that spends a lot of time on their feet.

Sample Day-to-Day Tasks:

- Examine, diagnose, and treat patients
- Prescribe medications and physical therapy
- Perform surgery
- Set fractures
- Fit patients for corrective shoes and orthotics
- Use and maintain high tech equipment
- Advise patients on preventative and curative care options

Work Environment:

Podiatrists typically work ~40 hour weeks in office environments. Interaction with the public can be rewarding or difficult depending on the client. Podiatrists can work in the military or civilian sectors, in clinics, hospitals, or private practice. Publication opportunities are high.

Required Training:

- Entry level requires a Doctor of Podiatric Medicine (D.P.M.) and a passing score on licensing examinations.
- After 3 or more years of study in a related field at the undergraduate level, a student may be admitted to a 4 year podiatry program.
- Most students go on to a 2-4 year residency program.
- All podiatrists must be licensed and pass a national exam.
- Continual learning through on the job training, journal articles, technical conferences, and classes is expected throughout one's career.

Employment Prospects/Pay:

With an aging and overweight population along with the increase in occurrences of diabetes, job growth in this area is expected to increase 23% over the next decade, about double the projected average for all occupations.

Salaries vary widely. Salaried or partnered podiatrists typically earn more and enjoy more opportunities than their private practice counterparts.

The median salary in 2012:
- $116,440

Mid-90-percentile range:
- $53,000 - $187,000

Radiation Therapist

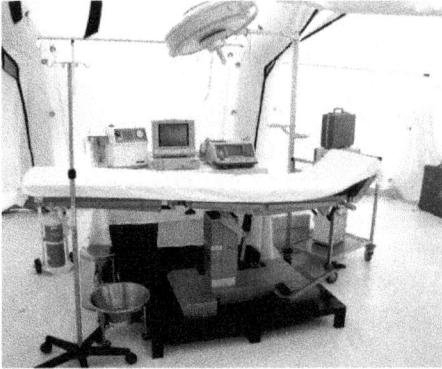

Description:

A radiation therapist treats cancer and other diseases with controlled doses of radiation. They work with machines called linear accelerators that deliver x-ray radiation to targeted areas of the body. They are responsible for the treatment and safety of their patients during the procedure. Anywhere from 1-4 years of study is required for this high-paying occupation.

Sample Day-to-Day Tasks:

- Counsel patients regarding condition and treatment options
- Conduct x-ray tests to locate specific areas for treatment
- Operate and maintain the linear accelerator
- Interface with computer systems to ensure accurate radiation delivery
- Follow safety protocols to protect themselves, staff, and patients
- Monitor patients for reactions to treatment
- Document patient treatment and progress

Work Environment:

Radiation therapists work ~40 hour weeks in hospital and clinic environments. They typically stand for long hours and need to be physically fit enough to help disabled patients. Interaction with the public can be rewarding or difficult depending on the client. Work is mainly found in the civilian sectors.

Required Training:

- A 12-month certification program is sufficient for some entry-level positions.
- An Associate's or Bachelor's degree in radiation therapy is preferred by most employers.
- Most states require licensure by passing a certification exam.

Employment Prospects/Pay:

Job growth is expected to increase 24% over the next decade, more than double the projected average for all occupations.

The median salary in 2012:
- $77,560

Mid-90-percentile range:
- $52,000 - $114,000

Top industries:
- Hospitals
- Physician's offices
- Outpatient centers

Radiologic or MRI Technologist

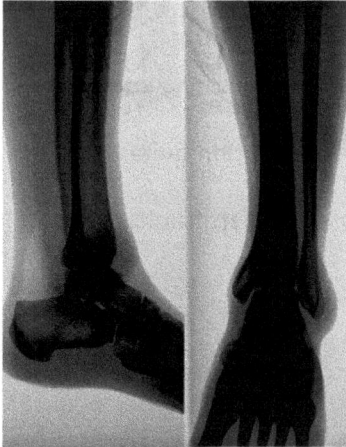

Description:

A radiologic technologist operates x-ray and other diagnostic equipment to determine patient illness. An MRI technologist operates a magnetic resonance imaging (MRI) scanner to do the same. Both technologists are responsible for the treatment and safety of their patients during the procedure. Only 2 years of study is required for this high-paying occupation.

Sample Day-to-Day Tasks:

- Counsel patients regarding questions about test procedures and what can be expected
- Conduct x-ray tests or MRI scans as per physicians' orders
- Operate and maintain equipment
- Interface with computer systems to ensure accurate operation of equipment
- Follow safety protocols to protect themselves, staff, and patients
- Monitor patients for reactions to treatment
- Document patient treatment and progress
- Assist is test result analysis

Work Environment:

Radiologic & MRI technologists work ~40 hour weeks in hospital and clinic environments. Shift work or evening and weekend work may be needed to cover emergency situations. On-call duties may be required, as well. They typically stand for long hours and need to be physically fit enough to help disabled patients. Interaction with the public can be rewarding or difficult depending on the client. Work is mainly found in the civilian sectors.

Required Training:

- An Associate's degree in radiography is the most common path for entry level work. A B.S. program is also available.
- Some states require licensure with passage of a certification exam.
- Many MRI technologists are first radiologic technologists.

Employment Prospects/Pay:

Job growth is expected to increase 21-24% over the next decade, about double the projected average for all occupations.

The median salary in 2012:
- $54,620 for radiologic technologists
- $65,360 for MRI technologists

Mid-90-percentile range:
- $37,000 - $77,000 for radiologic technologists
- $46,000 - $89,000 for MRI technologists

Top industries:
- Hospitals
- Physician's offices
- Medical laboratories

Registered Nurse

Description:

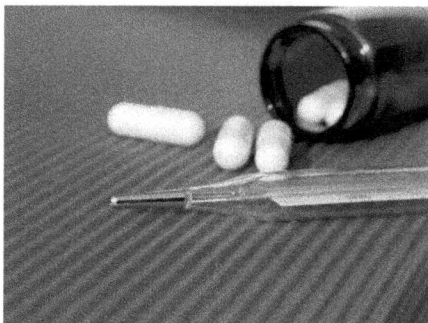

Registered nurses (RNs) are on the front lines of the health care industry. Many patients interact primarily with an RN. The registered nurse takes care of many aspects of patient care from documenting conditions to administering medications to counseling in treatment and preventative care. Intellect is just as important as patience and empathy in this demanding occupation.

Sample Day-to-Day Tasks:

- Record patient symptoms and histories
- Administer medications and treatments
- Counsel patients on conditions and treatment options
- Operate medical and diagnostic equipment
- Consult with physicians as needed
- Assist in test result analysis

Work Environment:

RNs work ~40 hour weeks in hospital, clinic, or home environments. Some positions require shift work or weekend and holiday work. Some RNs may also be on call depending on the employer and type of specialization. Willingness to relocate to areas without adequate medical care provides the best employment opportunities. The job can be physically and mentally demanding. Interaction with the public can be rewarding or difficult depending on the client. Work can be found in the military or civilian sectors.

Required Training:

- There are three possible educational paths to becoming an RN:
  - A B.S. in Nursing (BSN)
  - An Associate's Degree in Nursing (ADN)
  - A diploma from a nursing program.
- The BSN offers the best opportunities as it is required for administrative, research, consulting, and teaching positions. Also, some employers will only hire candidates with a BSN.
- RNs must become licensed by passing an national exam. Other requirements for licensure vary by state.
- Continual learning through on the job training, journal articles, technical conferences, and classes is expected throughout one's career.

Employment Prospects/Pay:

Job growth is expected to increase 19% over the next decade, about double the projected average for all occupations.

The median salary in 2012:
- $65,470

Mid-90-percentile range:
- $45,000 - $95,000

The average 2012 salary by top industries:
- $68,540 in government
- $67,210 in hospitals
- $62,090 in home care services
- $58,830 in nursing/care facilities
- $58,420 in physician's offices

Respiratory Therapist

Description:

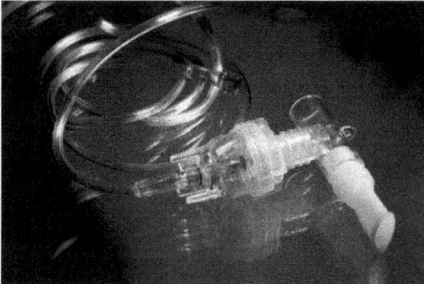

Respiratory therapists diagnose and treat breathing issues. They work with everyone from premature infants to the elderly. They deal with birth defects, underdeveloped lungs, and asthma as well as diseased lungs and emphysema. A two year degree is the minimum requirement for this important occupation.

Sample Day-to-Day Tasks:

- Examine patients to determine cause and severity of illness
- Counsel patients regarding condition and treatment options
- Conduct lung capacity and other diagnostic tests
- Monitor patients during testing
- Keep careful records of test process and results
- Prescribe and apply medicinal or physical therapies such as aerosol drugs or mucus remediation
- Instruct patients in proper home care
- Document patient treatment and progress
- Consult with other physicians

Work Environment:

Respiratory therapists work ~40 hour weeks in hospitals. Some work in care facilities or patient homes. Shift work and weekend and holiday work may be required.. They typically stand for long hours and need to be physically fit enough to help disabled patients. Interaction with the public can be rewarding or difficult depending on the client. Work is mainly found in the civilian sectors.

Required Training:

- An Associate's degree in respiratory therapy is the minimum requirement, but some employers prefer a B.S.
- Education is offered through universities, vocational institutes, and the armed forces. Academic as well as clinical work is a part of any accredited program.
- All states except Alaska require licensure. Requirements vary by state but usually involve passing a comprehensive exam.

Employment Prospects/Pay:

Job growth is expected to increase 19% over the next decade, about double the projected average for all occupations.

The median salary in 2012:
- $55,870

Mid-90-percentile range:
- $41,000 - $75,000

Top industries:
- Hospitals
- Nursing/care facilities
- Home care services

Speech & Language Pathologist

Description:

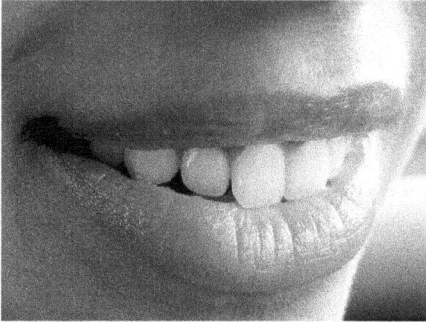

Speech & language pathologists are also called speech therapists. They help people who have suffered stroke, injury, disease, or trauma overcome physical impediments to speech and swallowing. They may treat a wide range of afflictions including physical disabilities, stuttering, and comprehension issues.

Sample Day-to-Day Tasks:

- Examine, test, diagnose, and treat patients
- Analyze severity of disabilities and impact on patient life
- Develop and implement treatment plans
- Teach alternative communications methods such as sign language as needed
- Help patients train the muscles required for speech and swallowing
- Help patients increase comprehension skills
- Counsel patients and their families
- Keep records of patient progress

Work Environment:

Speech & language pathologists typically work ~40 hour weeks. Patient interaction is a key component of the job, and local travel may be required of contractors. Interaction with the public can be rewarding or difficult depending on the client. Work is found in the civilian sector in schools or healthcare facilities. Publication opportunities are high.

Required Training:

- Entry level requires an M.S. in speech & language pathology.
- Supervised clinical work is part of the M.S. education.
- Licensing is required in most states.
- Additional certification may sometimes be required.
- Continual learning through on the job training, journal articles, technical conferences, and classes is expected throughout one's career.

Employment Prospects/Pay:

With an aging population, job growth in this area is expected to increase 19% over the next decade, about double the projected average for all occupations.

The median salary in 2012:
- $69,870

Mid-90-percentile range:
- $44,000 - $108,000

Top industries:
- K-12 schools
- Health practitioner offices
- Hospitals
- Nursing/care facilities

Veterinarian

Description:

Veterinarians get to heal our beloved four-legged friends. They can concentrate on small (e.g. dog, cat) or large (e.g. cow, horse) animals. They work in private practices, zoos, aquariums, and racetracks. There are many specialties in this field including dermatology, oncology, neurology, etc. Education and training runs the gamut of chemistry, biology, genetics, physiology, nutrition, physics, and math.

Sample Day-to-Day Tasks:

- Examine and treat animals
- Prescribe medication, set bones, give injections, collect samples as needed
- Euthanize
- Perform surgery
- Advise and console clients with sick animals
- Operate diagnostic equipment
- Perform administrative functions necessary to run a business
- Conduct research
- Supervise and train personnel

Work Environment:

Veterinarians typically work in private practices or on-site with larger animals. They tend to work long hours with a lot of standing. Noise and smells are part of the work atmosphere. Interaction with the public can be rewarding or difficult depending on the client. Veterinarians can work in the military or civilian sectors, in clinics, labs, zoos, aquariums, or racetracks. They can practice general medicine which includes office exams and surgery, or they can specialize. Research & publication is a path pursued by some. Management opportunities are available. Publication opportunities are high.

Required Training:

- Entry level requires a Doctor of Veterinary Medicine (D.V.M. or V.M.D.) and a state license.
- Students typically complete 4 years of undergraduate school in a biology, chemistry, or related field with 4 years of veterinary school afterward.
- Admission to veterinary schools is very competitive.
- Licensing varies by state, but all vets must pass an 8 hour national exam.
- Continual learning through on the job training, journal articles, technical conferences, and classes is expected throughout one's career.

Employment Prospects/Pay:

Job growth in this area is expected to increase 12% over the next decade, about the same as that projected average for all occupations.

The median salary in 2012:
- $84,460 overall
- $85,170 in federal government

Mid-90-percentile range:
- $52,000 - $144,000

Wages in 2011 by specialty:
- $71,096 in large animal, not equestrian
- $69,700 in companion animals
- $62,655 in mixed animal
- $43,404 in equestrian

Zoologist or Wildlife Specialist

Description:

Zoologists, or wildlife specialists, study the life processes of animals. Ornithologists specialize in birds, mammalogists in mammals, herpetologists in reptiles, and ichthyologists in fish. The purpose of a zoologist's work is to help in classification, study environmental impacts, learn about disease, or understand behaviors. Work with living or dead animals is an integral part of the job.

Sample Day-to-Day Tasks:

- Study animal behavior and health in their natural or simulated habitats
- Make recommendations about alternative industry practices where animals are adversely affected
- Dissect dead animals and analyze biological specimens
- Design and perform experiments
- Analyze data and predict trends
- Communicate results to management and the public in written or oral form
- Publish findings in scientific journals
- Supervise and train personnel

Work Environment:

Zoologists and wildlife specialists typically work 40 hour weeks in office and lab environments. Depending on the position, field work may be required which can involve travel, primitive living, and irregular hours. These biologists usually work in the civilian sector, typically in

academia. Most positions involve research or applied research.
Publication opportunities are high. Management opportunities exist.

Required Training:

- A B.S. or M.S. degree is the minimum requirement for many
  applied research jobs or K-12 teaching positions.
- Most researchers are required to have a Ph.D. Post-doctoral work
  and publications are an asset.
- Coursework involves chemistry, biology, mathematics, physics,
  engineering, and computer science.   These studies aid in running
  and creating automated tests and computer simulations.
- Continual learning through on the job training, journal articles,
  technical conferences, and classes is expected throughout one's
  career.

Employment Prospects/Pay:

Job growth in this area is expected to increase 5% over the next
decade, about half the projected average for all occupations.
Research positions and funding will be competitive.

The median salary in 2012:
- $57,710

Mid-90-percentile range:
- $37,000 - $95,000

The average 2012 salary by top industries:
- $72,700 in federal government
- $59,670 in research & development
- $57,710 in local government

Starting salaries for new biology baccalaureates in 2009:
- $33,254

Part 2: Technology

A field in which employment demands continue to grow exponentially. Technology includes anything you associate with computers. Examples of Technology careers include software engineer, network administrator (IT specialist), and computer scientist.

Computer & Information Research Scientist

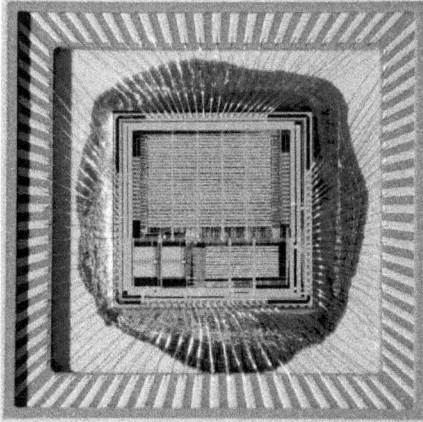

Description:

Computer & information research scientists (computer scientists) are the people who design and create new computer technologies. They are involved in the areas of computing hardware, robotics, and virtual reality. Creating new programming chips, developing robots for various applications, and designing games are a few of the projects computer scientists get to work on. They typically work closely with engineers of varying disciplines.

Sample Day-to-Day Tasks:

- Work with management to scope new projects
- Develop new computer technology in the field of computing hardware, robotics, or virtual reality
- Work in multi-disciplinary teams to create new technology products
- Formulate mathematical models for computer assisted solutions
- Supervise and train personnel
- Develop cost and schedule forecasts for programs

Work Environment:

Computer scientists typically work in offices with ~40 hour weeks. Some lab work is expected depending on the type of projects involved. Jobs are usually civilian in government, industry, or academia. Work is typically in the research and development area with opportunities for leadership positions, as well. Publication & patent opportunities are high.

Required Training:

- Most computer scientists have a Ph.D. in computer science or engineering with a strong math background.
- Federal government jobs sometimes have entry level positions for those with a B.S.
- Continual learning through on the job training, journal articles, technical conferences, and company funded classes is expected throughout one's career.

Employment Prospects/Pay:

With the boom of technological advancements, job growth is expected to increase by 15% over the next decade, faster than the average for all projected employment growth. Because many technical people who work with computers can find other technology careers with a B.S. or M.S., companies are currently scrambling to find qualified candidates with a Ph.D.

The median salary in 2012:
- $102,190

Mid-90-percentile range:
- $57,000 - $152,000

Top industries:
- Federal government
- Computer systems design services
- Academia
- Research & development
- Software publication

Computer & Information Systems Manager

Description:

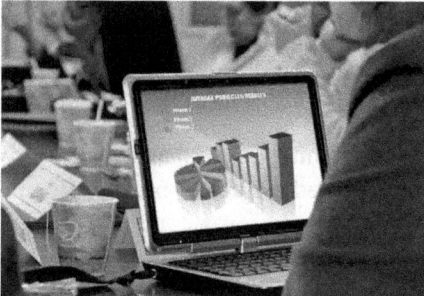

Computer and information systems managers direct other computer specialists including programmers, analysts, and information technology personnel. They are responsible for the implementation and administration of computer technology to aid business endeavors. Software development, network security, and internet usage are under their domain.

Sample Day-to-Day Tasks:

- Manage teams that maintain computer networks and systems
- Manage back-ups and recovery operations
- Provide technical support to diagnose and trouble-shoot problems
- Communicate closely with management to ensure optimal operation of systems
- Identify areas for improvement and growth
- Supervise, train, and hire personnel
- Develop cost and schedule forecasts for programs

Work Environment:

Computer and information systems managers typically work in offices with ~40 hour weeks, though a significant percentage work more with "on-call" duties required. Job stress can be high at times to handle unexpected problems or meet tight deadlines. Jobs can be found in military or civilian venues.

Required Training:

- Entry level positions require a B.S. in a computer related curriculum and years of relevant experience.
- Some positions strongly prefer an M.S. or M.B.A. along with relevant experience.
- Continual learning through on the job training, journal articles, technical conferences, and company funded classes is expected throughout one's career.

Employment Prospects/Pay:

Information technology and internet usage continue to grow at a rapid pace. Job growth is expected to increase by 15% over the next decade. The wages listed below do not include stock options and cash bonuses.

The median salary in 2012:
- $120,950

Mid-90-percentile range:
- $75,000 - >$187,000

The average 2012 salary by top industries:
- $133,120 in information
- $128,830 in computer systems design services
- $126,680 in finance & insurance
- $124,260 in management
- $101,690 in government

Computer Hardware Engineer

Description:

Computer hardware engineers design and create computing systems like desktops, laptops, and mobile devices as well as peripheral equipment like scanners, routers, and data storage devices. They design circuit boards, logic chips, and integrated systems. They work closely with software engineers to ensure new devices are compatible with new operating systems and programs.

Sample Day-to-Day Tasks:

- Develop new computer hardware technologies from chip to system level
- Consult with software developers
- Test new hardware for functionality and software compatibility
- Develop manufacturing techniques
- Oversee production
- Supervise and train personnel
- Develop cost and schedule forecasts for programs

Work Environment:

Computer hardware engineers typically work in labs and offices with ~40 hour weeks, though about one quarter of them work additional hours. Jobs are usually civilian in government and private industry, in manufacturing and research and development. There are opportunities for leadership positions with experience. Publication & patent opportunities are high.

Required Training:

- Most positions require a B.S. in computer or electrical engineering.
- Some positions require an M.S. or M.B.A. in addition to undergraduate study.
- Experience in programming is usually required.
- Continual learning through on the job training, journal articles, technical conferences, and company funded classes is expected throughout one's career.

Employment Prospects/Pay:

Job growth is expected to increase by 7% over the next decade, a little slower than the average for all projected employment growth.

The median salary in 2012:
- $100,920

Mid-90-percentile range:
- $64,000 - $150,000

The average 2012 salary by top industries:
- $109,860 in computer & semiconductor manufacturing
- $101,680 in component manufacturing
- $101,510 in computer systems design services
- $97,970 in various instrumentation industries
- $95,710 in research & development

Computer Programmer

Description:

Computer programmers turn the algorithms and flowcharts of computer software engineers into executable code. They write programs for computers, mobile systems, and internet usage. They update software to be compatible with various operating systems and debug programs as needed.

Sample Day-to-Day Tasks:

- Interpret software engineers' designs and convert them to executable code
- Some programmers take on software engineering responsibilities in scoping and designing what the program must do
- Test and debug code to ensure proper operation
- Rewrite code for various operating systems
- Expand functionality of existing programs
- Use computer assisted tools to develop code

Work Environment:

Computer programmers typically work in offices with ~40 hour weeks, though telecommuting is becoming more common. Most work is done by a single individual, but team work is required of larger projects. Jobs are usually civilian in government, industry, or academia. Opportunities for advancement and leadership are available.

Required Training:

- Most computer programmers have a B.S. in computer science or engineering.
- Some positions will hire candidates with an Associate's Degree.
- A few languages are chosen as specialties with other languages learned on the job.
- Experience is highly valued by employers, so internships are common.
- Certification in specific languages or software systems is sometimes obtained to make a programmer more marketable.
- Continual learning through on the job training, journal articles, technical conferences, and company funded classes is expected throughout one's career.

Employment Prospects/Pay:

Job growth is expected to increase by 8% over the next decade, about the average for all projected employment growth. Outsourcing is an issue with U.S. jobs in this field, but small companies and even some large companies still prefer to work with local candidates.

The median salary in 2012:
- $74,280

Mid-90-percentile range:
- $43,000 - $118,000

Computer Software Engineer/Developer

Description:

Computer software engineers are also known as software developers The programs they develop are used by regular consumers, government, and businesses and include utility software, application packages, and automation of business operations to name a very few. Software developed by computer engineers is then passed to programmers who code the instructions into an appropriate computing language.

Sample Day-to-Day Tasks:

- Develop new software algorithms for various applications
- Determine hardware and power requirements for new projects
- Consult with customers to develop customized solutions
- Direct programming and documentation efforts
- Develop testing and validation procedures
- Supervise and train personnel
- Develop cost and schedule forecasts for programs

Work Environment:

Computer software engineers typically work in offices with ~40 hour weeks. Some travel to customer locations is expected depending on the type of projects involved. Jobs are usually civilian in government, industry, or academia. Work is typically in the research and development area with opportunities for leadership positions, as well. Publication & patent opportunities are high.

Required Training:

- Entry level positions require a B.S. in a computer related curriculum or mathematics.
- An M.S. degree is sometimes needed.
- A high value is placed on broad experience with different platforms and languages.
- Continual learning through on the job training, journal articles, technical conferences, and company funded classes is expected throughout one's career.

Employment Prospects/Pay:

The computer science area continues to grow at a rapid pace. Job growth is expected to increase by 22% over the next decade, about double the average for all projected employment growth.

The median salary in 2012:
- $90,060 for applications software developers
- $99,000 for systems software developers

Mid-90-percentile range:
- $55,000 - $139,000 for applications software developers
- $63,000 - $149,000 for systems software developers

The average 2012 salary by top industries for applications SW Dev.:
- $97,960 in computer product manufacturing
- $96,920 in software publication
- $91,970 in finance & insurance
- $88,500 in computer systems design services

The average 2012 salary by top industries for systems SW Dev.:
- $105,030 in computer product manufacturing
- $99,940 in finance & insurance
- $99,750 in software publication
- $98,500 in computer systems design services

Computer Systems Analyst

Description:

Computer systems analysts help businesses use computers to streamline their processes with new or existing technologies. Analysts configure hardware and software and fine-tune existing systems to maximize efficiency for the customer. Programming, systems evaluation, and network administration are an integral part of this career.

Sample Day-to-Day Tasks:

- Modify current systems to increase productivity and utility for the customer
- Develop test procedures and quality assurance standards
- Assist in problem solving and trouble-shooting
- Develop flow diagrams and code to customize software systems
- Coordinate communications and network activities
- Consult with customers to develop customized solutions
- Conduct programming and documentation efforts
- Supervise and train personnel
- Develop cost and schedule forecasts for programs

Work Environment:

Computer systems analysts typically work in offices with ~40 hour weeks. Telecommuting full-time is an option in some cases. Jobs are usually civilian in government, industry, or academia. Work is typically in the research and development area with opportunities for leadership positions, as well. Publication & patent opportunities are high.

Required Training:

- Entry level positions usually require a B.S. in a computer related curriculum.
- A technical graduate degree or M.B.A. is sometimes preferred.
- Relevant experience is a definite plus.
- Often computer systems analysts are former computer programmers.
- Continual learning through on the job training, journal articles, technical conferences, and company funded classes is expected throughout one's career.

Employment Prospects/Pay:

Information technology and internet usage continue to grow at a rapid pace. Job growth is expected to increase by 25% over the next decade, more than double the average for all projected employment growth.

The median salary in 2012:
- $79,680

Mid-90-percentile range:
- $50,000 - $122,000

Top industries:
- Computer systems design services
- Finance & insurance
- Management
- Information
- State & local government

Database Administrator

Description:

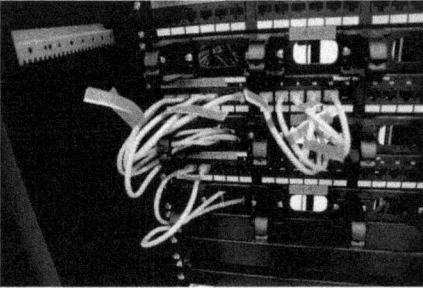

Database administrators (DBA's) ensure safe and secure storage of information. They develop database systems and test them for functionality, security, and integrity. Specializations include system (general) and application (specific to a certain functionality). Paths to this career are somewhat flexible.

Sample Day-to-Day Tasks:

- Scope customer database requirements for amount of data storage, ease of use, and security requirements
- Develop database software to store information
- Train customers on interfacing with the database
- Test the database for functionality, security, and integrity
- Maintain and fix database systems as needed
- Perform back-ups and expansions
- Install upgrades and patches

Work Environment:

Database administrators typically work in offices with ~40 hour weeks, though about a quarter work more hours. Job stress can be high at times to handle unexpected problems or meet tight deadlines. Jobs are usually civilian in banking, insurance, information, and educational institutions. Management opportunities are available.

Required Training:

- Entry level positions require a B.S. in a computer related field.
- Some positions require an M.B.A.
- Most database administrator positions are only open to candidates who have successfully performed in other computer related careers such as database developer or analyst.
- A thorough knowledge of SQL or similar is required.
- Continual learning through on the job training, journal articles, technical conferences, and company funded classes is expected throughout one's career.

Employment Prospects/Pay:

Job growth is expected to increase by 15% over the next decade, faster than the average for all projected employment growth.

The median salary in 2012:
- $77,080

Mid-90-percentile range:
- $43,000 - $119,000

The average 2012 salary by top industries:
- $85,880 in finance & insurance
- $84,550 in computer systems design services
- $82,290 in management
- $81,800 in information
- $63,620 in education services

Network & Systems Administrator

Description:

Network and computer systems administrators are the information technology (IT) specialists of the business world. They keep the computer systems running and ensure the networks are configured and working properly. Education paths to this career are somewhat flexible.

Sample Day-to-Day Tasks:

- Maintain computer networks and systems
- Perform back-ups and recovery operations
- Diagnose and trouble-shoot problems related to hardware and software
- Plan and implement security and virus protection systems
- Configure and monitor networks
- Maintain inventories and keep careful records
- Supervise and train personnel
- Develop cost and schedule forecasts for programs

Work Environment:

Network and computer systems administrators typically work in offices with ~40 hour weeks, though a significant percentage work more "on-call" duties. Job stress can be high at times to handle unexpected problems or meet tight deadlines. Interfacing with customers or clients who have "emergencies" can also be stressful. Jobs are usually civilian in government, industry, or academia. Management opportunities are available.

Required Training:

- Entry level positions require an associate's degree or professional certification, though many jobs prefer a B.S. in a computer related curriculum.
- A strong emphasis is placed on relevant experience.
- Continual learning through on the job training, journal articles, technical conferences, and company funded classes is expected throughout one's career.

Employment Prospects/Pay:

Job growth is expected to increase by 12% over the next decade, about the average for all projected employment growth.

The median salary in 2012:
- $72,560

Mid-90-percentile range:
- $44,000 - $115,000

The average 2012 salary by top industries:
- $77,370 in finance & insurance
- $77,270 in information
- $76,090 in computer systems design services
- $70,250 in manufacturing
- $61,830 in education services

Security Analyst or Network Architect

Description:

Information security analysts ensure an organizations systems are not vulnerable to outside attack. Computer network architects develop the systems that an organization's employees use to communicate and work with each other. Both are vital in today's economy as the need for computer-assisted, secure communications grows.

Sample Day-to-Day Tasks:

- Stay on top of new advances in the field
- Develop security protocols and monitor systems for violations
- Install firewall s and data encryption programs
- Develop hardware and software requirements for a network
- Design cable layouts and hardware placement
- Install new system architectures
- Work with management to implement security or systems architecture plans
- Train personnel in proper usage of security and network systems
- Develop cost and schedule forecasts for programs

Work Environment:

Information security analysts and computer network architects typically work in offices with ~40 hour weeks. Sometimes security analysts may need to work extra hours or "on call" to deal with emergencies. Job stress can be high when such emergencies occur. Jobs are usually civilian in government, industry, or academia. Management opportunities are available.

Required Training:

- Entry level positions require a B.S. in computer science or engineering.
- Some employers also require an M.B.A.
- Most positions are only open to candidates who have successfully performed in other computer related careers for 5-10 years.
- Continual learning through on the job training, journal articles, technical conferences, and company funded classes is expected throughout one's career.

Employment Prospects/Pay:

Information technology and internet usage continue to grow at a rapid pace. Job growth for security analysts is expected to increase by 37% over the next decade, over triple the average for all projected employment growth. In the same period, job growth is expected to be 15% for network architects, still faster than average.

The median salary in 2012:
- $86,170 for security analysts
- $91,000 for network architects

Mid-90-percentile range:
- $50,000 - $136,000 for security analysts
- $53,000 - $142,000 for network architects

Top industries:
- Computer systems designs services
- Finance & insurance
- Management
- Telecommunications & information

Web Developer

Description:

Web developers, no surprise, develop websites. They need to have creativity in developing the visual aspects of a web interface, and they need to be technically proficient in order to deliver functionality. Websites can simply convey information, but most are interactive and require knowledge of programming languages beyond html.

Sample Day-to-Day Tasks:

- Conduct client meetings to determine needs in aesthetics, functionality, and capacity
- Use programming languages to develop interfaces, interactive components, databases, or other as needed
- Integrate various components of a site to work seamlessly together
- Work with graphic designers as needed
- Work with members of a team for larger projects
- Debug issues with the site
- Instruct clients on how to use and maintain the site
- Monitor site statistics like traffic to the site

Work Environment:

About 25% of web developers work for themselves and set their own hours. The rest are employed in offices with ~40 hour weeks. Job stress can be high when short deadlines occur. Jobs are usually found in the civilian sector. Management positions are available for those with a B.S.

Required Training:

- Most web developers have an Associate's degree in web design. However, demonstrative experience is sufficient for some employers.
- Some web developers have a B.S. in computer science for the more technical positions.
- A variety of programming languages as well as graphic design capability is an asset.
- Continual learning is expected throughout one's career.

Employment Prospects/Pay:

Information technology and internet usage continue to grow at a rapid pace. Job growth is expected to increase by 20% over the next decade, about double the average for all projected employment growth.

The median salary in 2012:
- $62,500

Mid-90-percentile range:
- $34,000 - $105,000

Top industries
- Computer system design services
- Information
- Finance & insurance
- Education services
- Religious & civic services

Part 3: Engineering

Another very broad category. Engineering runs the gamut from electrical and mechanical to civil and industrial and more. Examples of Engineering careers include materials engineer, nuclear engineer, and architect.

Aerospace Engineer

Description:

Aerospace engineers design and build machines and vessels that travel through air or space. Specifically, they work on structure, navigation control, and instrumentation for commercial airplanes and helicopters, fighter jets and drones, space shuttles and satellites, as well as missiles and rockets. Disciplines include aerodynamics, thermodynamics, celestial mechanics, propulsion, acoustics, and guidance and control systems.

Sample Day-to-Day Tasks:

- Use computer assisted design tools to prepare technical drawings
- Perform detailed calculations of operating parameters and specifications for a given project
- Design structure, navigation controls, or instrumentation for air or space systems
- Design and perform various experiments to test designs and equipment compliance
- Coordinate investigation of technical issues with customer equipment or research demonstrations
- Perform inspections to ensure compliance with safety and operation protocols
- Work with other engineers and customers to provide customized solutions
- Supervise and train personnel
- Develop cost and schedule forecasts for programs

Work Environment:

Engineers typically work in offices with ~40 hour weeks. Sometimes engineers are required to travel to perform on-site installations or confer with customers. Desk work as well as lab work is expected, though the mix varies by job title and personal preference. Jobs can be found in both military and civilian venues. Engineers can work in the research, development, or manufacturing areas. Engineers can also perform as technical contributors, program managers, or functional managers and supervisors. Opportunities for publications and patents are high.

Required Training:

- Entry level requires a B.S. in a related engineering discipline.
- A graduate of a four year technology program may qualify for entry level positions, as well, but cannot obtain a professional engineering (P.E.) license. P.E. certification is required to work directly with the public but is not required for most industry jobs.
- Research and development professionals as well as college faculty in engineering are required to have an M.S. or usually a Ph.D.
- Continual learning through on the job training, journal articles, technical conferences, and company funded classes is expected throughout one's career.

Employment Prospects/Pay:

With cuts in NASA and government funding, the space and defense industries are expected to grow 7% over the next decade, a little slower than the average of all occupations. Retirement trends and an extreme lack of eligible engineers with U.S. citizenship will help drive the need for new recruits.

The median salary in 2012:
- $103,720 ($71,859 for entry level)

Mid-90-percentile range:
- $65,000 - $149,000

Starting salaries for new aerospace engineering baccalaureates in 2009:
- $56,311

Aircraft Mechanic or Avionics Technician

Description:

Aircraft mechanics maintain and inspect the mechanical and electrical systems of aviation equipment. Avionics technicians maintain and inspect the electronics and instrumentation systems. Both work with state of the art equipment to fix problems and ensure safe transport.

Sample Day-to-Day Tasks:

- Inspect aircraft for damage or faults
- Perform maintenance such as oil changes and function checks
- Repair or replace mechanical or electrical parts as needed
- Assemble sub-components to larger systems
- Use oscilloscopes and circuit testers to diagnose equipment
- Interpret flight and test data
- Keep detailed records

Work Environment:

Aircraft mechanics and avionics technicians typically work in the field with ~40 hour weeks. Rotating shifts as well as weekend and holiday work are common. Jobs can be found in both military and civilian venues. The work can be dangerous owing to working with heavy machinery. Physical agility is important as lifting, standing, and kneeling are regular components of the job. Noise as well as hot and cold temperatures are a common part of the work environment.

Required Training:

- There are three avenues to employment: on-the-job training after high school, military training, or graduation from an FAA-approved Aviation Maintenance program.
- Certification is encouraged for advancement and best job prospects.

Employment Prospects/Pay:

Job growth is expected to be 2% over the next decade, slower than the average of all occupations.

The median salary in 2012:
- $55,210 for aircraft mechanics
- $55,350 for avionics technicians

Mid-90-percentile range:
- $35,000 - $77,000 for aircraft mechanics
- $39,000 - $74,000 for avionics technicians

The average 2012 salary by top industries for aircraft mechanics:
- $59,110 in air transportation
- $55,940 in federal government
- $55,650 in aerospace manufacturing

The average 2012 salary by top industries for avionics technicians:
- $60,780 in aerospace manufacturing
- $59,750 in professional & technical services
- $58,530 in air transportation

Agricultural Engineer

Description:

Agricultural engineering encompasses the development of processes and equipment associated with farming. From soil and water conservation to sensor design, from innovations in machinery and structures to improvements in agriculture processing techniques, these engineers strive to improve the quality and efficiency of our food supply chain. They find work in research and development, manufacturing, and marketing.

Sample Day-to-Day Tasks:

- Use computer assisted tools to design machine components and equipment
- Design and coordinate land reclamation projects, food processing plants, and structures for storage, shelter, loading, and processing
- Monitor and advise on water quality, pollution control, irrigation sources, and electrical distribution systems
- Prepare technical drawings
- Perform inspections to ensure compliance with safety protocols and regulations
- Work with other engineers and customers to provide customized solutions
- Develop cost and schedule forecasts for programs

Work Environment:

Engineers typically work in offices with ~40 hour weeks. Sometimes engineers are required to travel to perform on-site installations or confer with customers. Jobs can be found usually in the civilian sector. Engineers can work in the research, development, manufacturing, or marketing areas. Engineers can also perform as technical contributors, program managers, or functional managers and supervisors.

Required Training:

- Entry level requires a B.S. in a related engineering discipline.
- A graduate of a four year technology program may qualify for entry level positions, as well, but cannot obtain a professional engineering (P.E.) license. P.E. certification is required to work directly with the public but is not required for most industry jobs.
- Research and development professionals as well as college faculty in engineering are required to have an M.S. or usually a Ph.D.
- Continual learning through on the job training, journal articles, technical conferences, and classes is expected throughout one's career.

Employment Prospects/Pay:

Even with a growing population as well as an increased demand for biofuels, job growth in this area is expected to increase by 5% over the next decade, which is about half the average for all occupations.

The median salary in 2012:
- $74,000

Mid-90-percentile range:
- $45,000 - $116,000

The average 2012 salary by top industries:
- $82,090 in engineering services
- $77,030 in federal government
- $73,380 in food manufacturing
- $67,690 in machinery manufacturing
- $50,100 in education services

Starting salaries for new agricultural engineering baccalaureates in 2009:
- $54,352

Architect

Description:

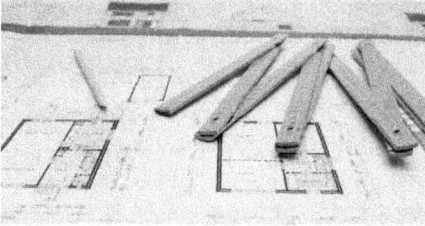

Architects design buildings and structures for commercial, industrial, public, and personal use. They find the balance between budget, schedule, functionality, safety, and aesthetic beauty. Architects can see a project through from conceptualization to final build and inspection. Engineering, communications, and management skills are needed along with design talent.

Sample Day-to-Day Tasks:

- Consult with clients to assess their needs
- Plan layouts and develop scaled drawings
- Engineer structures for safety and compliance with regulation
- Compile requirements for equipment, materials, and labor
- Supervise and train personnel to help with drawings and documentation
- Prepare and administer contractual documentation
- Perform on-site inspections of construction sites
- Generate new work through proposals, marketing, and presentations

Work Environment:

Architects typically work in offices with ~40 hour weeks. Sometimes architects visit construction sites, and deadlines can require some overtime work. Jobs can be found in both military and civilian venues. Approximately 20% of architects are self-employed which is about triple that of all other occupations.

Required Training:

- Three steps are required to become a practicing architect.
  1.) The education requirement is completion of a 5 year bachelor of architecture program, or, if a B.S. is held in another field, an M.S. in architecture can be undertaken. The M.S. can last 1-5 years depending on the extent of prior architecture study. As a word of caution, a B.Arch. is dissimilar enough to a B.S. in other curricula that deciding to transfer out of architecture mid-way may be equivalent to starting college over.
  2.) An internship of at least 3 years is required.
  3.) Finally, a candidate must pass all 9 portions of the Architect Registration Examination.
- Continual learning through on the job training, technical conferences, and company funded classes is expected throughout one's career.

Employment Prospects/Pay:

Job growth is expected to rise 17% over the next decade, about double the average for all other occupations. Competition is expected to be high for the most prestigious firms.

Internships pay considerably less, and partners in prestigious firms can make considerably more.

The median salary in 2012:
- $73,090

Mid-90-percentile range:
- $45,000 - $118,000

Biomedical Engineer

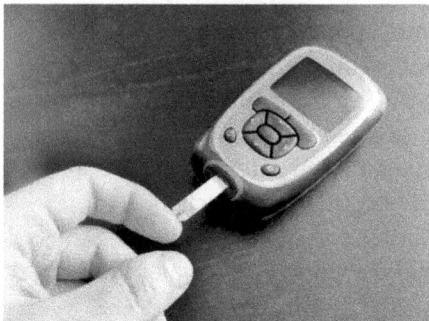

Description:

Biomedical engineers design and build devices associated with the medical industry. They help develop artificial organs and prosthetics as well as instruments like pace-makers, sleep APNEA monitors, insulin injection devices, and biomedical imaging systems. Disciplines include electrical and mechanical engineering specialties as well as strong biomedical training.

Sample Day-to-Day Tasks:

- Develop new medical instrumentation or test equipment
- Conduct research to gain insight on the engineering side of biological systems
- Evaluate designs and equipment for ease of use, proper operation, and safety
- Interpret data to develop new devices and forecast effectiveness
- Research new materials for bio-compatibility
- Conduct training for customers and patients
- Work with other engineers and medical specialists to provide innovative solutions
- Supervise and train personnel
- Develop cost and schedule forecasts for programs

Work Environment:

Engineers typically work in offices with ~40 hour weeks. Sometimes engineers are required to travel to perform on-site installations or confer with customers. Desk work as well as lab work is expected, though the mix varies by job title and personal preference. Jobs can be found in both military and civilian venues. Engineers can work in the research, development, or manufacturing areas. Engineers can

Required Training:

- Entry level requires a B.S. in a related engineering discipline, but a graduate degree is strongly encouraged.
- Research and development professionals as well as college faculty in engineering are required to have an M.S. or usually a Ph.D.
- Continual learning through on the job training, journal articles, technical conferences, and company funded classes is expected throughout one's career.

Employment Prospects/Pay:

With an aging population and growing health-care concerns, job growth is expected to increase by 27% over the next decade, almost triple that of the average career.

The median salary in 2012:
- $86,960

Mid-90-percentile range:
- $53,000 - $139,000

The average 2012 salary by top industries:
- $94,150 in research & development
- $88,850 in medical manufacturing
- $87,340 in pharmaceutical manufacturing
- $69,910 in hospitals
- $63,440 in academia

Starting salaries for new biomedical engineering baccalaureates in 2009:
- $54,158

Cartographer or Photogrammetrist

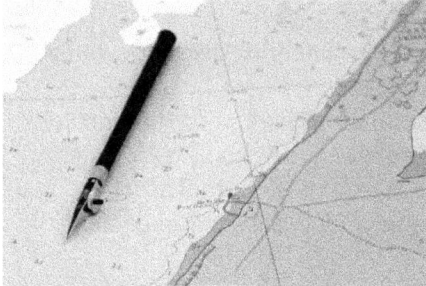

Description:

A cartographer or photogrammetrist collects data about the surface and climate of the Earth. They compile their data into maps and other visual representations. A cartographer specializes in map creation whereas a photogrammetrist uses imagery to create surface relief models.

Sample Day-to-Day Tasks:

- Use computer assisted tools to collect data
- Compile information about precipitation levels, climate, surface topology, and landmarks
- Travel to take measurements in the field
- Interpret data from imaging sensors and light detection systems
- Plan surveys and data collection efforts
- Assemble data into maps and other visual formats
- Update and revise existing information

Work Environment:

Cartographers and photogrammetrists typically work in offices with ~40 hour weeks. Extensive travel to areas being investigated is sometimes required and result in longer work days. Work is typically found in the civilian sector with architectural and engineering firms as well as in various levels of government.

Required Training:

- Entry level requires a B.S. in cartography, geography, civil engineering, or a related discipline.
- Some states require licensing as a surveyor.
- Some states have additional licensing requirements for photogrammetrists.

Employment Prospects/Pay:

Job growth is expected to be 20% over the next decade, about double the average of all occupations.

The median salary in 2012:
- $57,440

Mid-90-percentile range:
- $35,000 - $95,000

The average 2012 salary by top industries:
- $84,850 in federal government
- $57,780 in local government
- $57,180 in consulting
- $55,260 in architectural services
- $51,910 in state government

Chemical Engineer

Description:

Chemical engineering combines chemistry with industry. Chemical engineers handle large scale chemical production and by-product waste management. They are involved in anything from commercial products such as clothing and paper to biomedical products and instrumentation. Areas of study in addition to chemistry include physics, mathematics, mechanical engineering, and electrical engineering.

Sample Day-to-Day Tasks:

- Develop new chemical manufacturing processes and handling techniques
- Analyze worker and environmental safety with respect to chemical usage and storage
- Evaluate manufacturing processes for consistency in fabrication, safety, and government compliance
- Design factory layouts and trouble-shoot problems
- Develop safety protocols and train personnel as needed
- Work with engineers and chemists to provide customized solutions
- Develop cost and schedule forecasts for programs

Work Environment:

Engineers typically work in offices with ~40 hour weeks. Sometimes engineers are required to travel to perform on-site installations or confer with customers. Desk work as well as lab work is expected, though the mix varies by job title and personal preference. Jobs can be found in both military and civilian venues. Engineers can work in the research, development, or manufacturing areas. Engineers can also perform as technical contributors, program managers, or functional managers and supervisors. Opportunities for publications and patents are high.

Required Training:

- Entry level requires a B.S. in a related engineering discipline.
- A graduate of a four year technology program may qualify for entry level positions, as well, but cannot obtain a professional engineering (P.E.) license. P.E. certification is required to work directly with the public but is not required for most industry jobs.
- Research and development professionals as well as college faculty in engineering are required to have an M.S. or usually a Ph.D.
- Continual learning through on the job training, journal articles, technical conferences, and company funded classes is expected throughout one's career.

Employment Prospects/Pay:

Chemical engineering jobs are expected grow 4% over the next decade, about half the average of all other occupations.

The median salary in 2012:
- $94,350

Mid-90-percentile range:
- $59,000 - $155,000

The average 2012 salary by top industries:
- $105,310 in petroleum manufacturing
- $99,510 in chemical manufacturing
- $97,880 in research & development

Starting salaries for new chemical engineering baccalaureates in 2009:
- $64,902

Civil Engineer

Description:

Civil engineers are responsible for the design and build of bridges, sewer systems, dams, tunnels, airports, and anything else that makes our civilization run. Specialties include structural design, water studies, construction, transportation, and geotechnical engineering. Most positions are supervisory in nature.

Sample Day-to-Day Tasks:

- Use computer assisted design tools to prepare technical drawings
- Manage teams to design and implement large structural projects
- Perform detailed calculations to determine safety of a particular design
- Perform experiments to determine site suitability for a particular installation
- Analyze reports, maps, photos, and drawings to plan projects
- Analyze statistical data to propose new solutions to problems
- Present and defend public reports and proposals
- Provide technical guidance and training
- Perform inspections to ensure compliance with safety and operation protocols
- Develop cost and schedule forecasts for programs

Work Environment:

Engineers typically work in offices with ~40 hour weeks. Desk work as well as on-site work is expected, though the mix varies by job title and personal preference. Jobs can be found in both military and civilian venues. Civil engineers work in private construction or as a city engineer with supervisory or management responsibilities. Opportunities for publications and patents are high.

Required Training:

- Entry level requires a B.S. in a related engineering discipline. Cooperative education experience is a great asset.
- A graduate of a four year technology program may qualify for entry level positions, as well, but cannot obtain a professional engineering (P.E.) license. P.E. certification is required to work directly with the public which may be required of civil engineers.
- Research and development professionals as well as college faculty in engineering are required to have an M.S. or usually a Ph.D.
- Continual learning through on the job training, journal articles, technical conferences, and company funded classes is expected throughout one's career.

Employment Prospects/Pay:

Despite cuts in general construction, the country's ailing infrastructure, ever-growing population, and change in government regulations will keep the demand for civil engineers high. A growth of 20% is expected over the next decade which is about double the average expected job growth across all industries.

The median salary in 2012:
- $79,340

Mid-90-percentile range:
- $51,000 - $122,000

The average 2012 salary by top industries:
- $89,440 in federal government
- $83,670 in local government
- $79,470 in architectural services
- $74,180 in state government

Starting salaries for new civil engineering baccalaureates in 2009:
- $52,048

Electrical Engineer

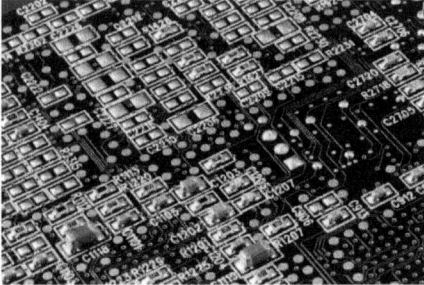

Description:

Electrical engineering covers a very broad range of occupational areas. Concentrations within the major include electronics, power systems, electro-magnetic waves (think lasers and optics), signal processing and communication systems, and computer hardware engineering. An electrical engineer is responsible for design and implementation of systems within their specialty. They can be involved in anything from the electronics for a cool new gadget to major power distribution systems to medical or scientific instrumentation to new communications encoding algorithms.

Sample Day-to-Day Tasks:

- Use computer assisted design tools to prepare technical drawings
- Perform detailed calculations of operating parameters and specifications for a given project
- Design and implement innovative electrical systems
- Coordinate and direct technicians or manufacturing staff in the implementation of designs
- Perform inspections to ensure compliance with safety and operation protocols
- Work with other engineers and customers to provide customized solutions
- Supervise and train personnel
- Develop cost and schedule forecasts for programs

Work Environment:

Engineers typically work in offices with ~40 hour weeks. Sometimes engineers are required to travel to perform on-site installations or confer with customers. Desk work as well as lab work is expected, though the mix varies by job title and personal preference. Jobs can be found in both military and civilian venues. Engineers can work in

the research, development, or manufacturing areas. Engineers can also perform as technical contributors, program managers, or functional managers and supervisors. Opportunities for publications and patents are high.

Required Training:

- Entry level requires a B.S. in a related engineering discipline.
- A graduate of a four year technology program may qualify for entry level positions, as well, but cannot obtain a professional engineering (P.E.) license. P.E. certification is required to work directly with the public but is not required for most industry jobs.
- Research and development professionals as well as college faculty in engineering are required to have an M.S. or usually a Ph.D.
- Continual learning through on the job training, journal articles, technical conferences, and company funded classes is expected throughout one's career.

Employment Prospects/Pay:

Though the consumer demand for electronics goods is expected to rise, job growth in this area is expected to increase by only 4% owing to foreign competition. Electrical engineers do enjoy opportunities in a variety of other fields including aerospace, optical science, and computer technology. Top employers include the federal government and its contractors (which are desperate for eligible U.S. citizens) and semiconductor manufacturing.

The median salary in 2012:
- $89,630

Mid-90-percentile range:
- $58,000 - $141,000

Starting salaries for new electrical engineering baccalaureates in 2009:
- $60,125

Engineering Manager

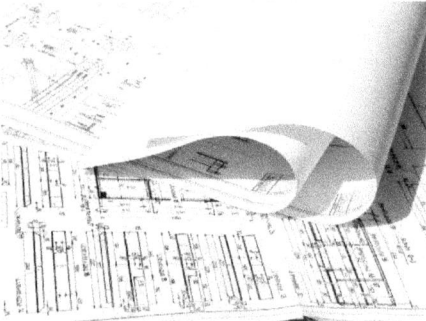

Description:

Engineering managers oversee teams of scientists and engineers. They help come up with programs to meet executive goals, ensure the projects are adequately staffed, and verify that employee concerns are addressed. They may take on a more project oriented role in directing activities and taking responsibility for technical, fiscal, and schedule successes. Or, they may operate from a functional standpoint of bridging the gap between staff and executives and having regular career related discussions with employees. In order to be effective, engineering managers must have a strong background and solid experience in the field they oversee.

Sample Day-to-Day Tasks:

- Develop programs to meet broad goals set by exec. management
- Act as communications bridge between executives and staff
- Develop detailed project plans to meet technical, fiscal, and schedule goals
- Ensure staffing is sufficient on programs and make sure all employees are being utilized equally
- Make presentations for internal and external consumption
- Conduct performance reviews
- Give and solicit regular feedback to and from staff
- Recruit, hire, and fire as required
- Enforce company regulations and set an example of compliance
- Approve time cards, prepare reports, and conduct other administrative duties

Work Environment:

Engineering managers typically work in offices with ~40 hour weeks, though overtime work can be frequent. Some managers are required to travel to perform on-site inspections or to confer with customers.

Jobs can be found in both military and civilian venues. Engineering managers can work in the research, development, or manufacturing areas. Some also continue to do technical work while others find that their work is limited to management responsibilities.

Required Training:

- Entry level requires a B.S. in a related engineering or science discipline along with years of experience as a technical contributor in that field.
- Many engineering managers also get a Masters of Business Administration (M.B.A.) or a Masters of Engineering Management (MEM). Many have advanced degrees in their field instead of or in addition to a business or management degree.
- The education and experience requirements will vary by company, but a strong technical background along with an ability to manage people and translate technical jargon to management are required demonstrated skills.
- Continual learning through on the job training, journal articles, technical conferences, and company funded classes is expected throughout one's career.

Employment Prospects/Pay:

Job growth is expected to rise 7% over the next decade, slightly slower than average. The wages listed below do not include stock options and cash bonuses.

The median salary in 2012:
- $124,870

Mid-90-percentile range:
- $80,000 - >$187,000

The average 2012 salary by top industries:
- $147,250 in mining, oil, & gas
- $142,310 in research & development
- $124,000 in manufacturing

Engineering Technician

Description:

Engineering technicians assist engineers in laboratory environments. They carry out experiments, document results, and assist in planning projects. Their work is very hands-on, and their efforts directly contribute to the success of an experiment or the creation of a new prototype.
A two year technical education is required for this type of work. Technicians train to specialize in one of the engineering disciplines.

Sample Day-to-Day Tasks:

- Use computer assisted design tools to prepare technical drawings
- Carry out experiments alone or in teams
- Provide detailed documentation of results in laboratory notebooks or written reports
- Ensure safe laboratory practices
- Work with engineers to help plan future tasks
- Coordinate with colleagues on larger efforts

Work Environment:

Engineering technicians typically work in offices with ~40 hour weeks. Desk work as well as lab work is expected with emphasis on the latter. Sometimes technicians are required to travel to perform on-site installations or confer with customers. Jobs can be found in both military and civilian venues. Technicians typically work in work in the research, development, or manufacturing areas. Opportunities to contribute to publications and patents are high.

Required Training:

- Entry level requires an Associate's degree in a related field of study.
- Continual learning through on the job training and company funded classes is expected throughout one's career.

Employment Prospects/Pay:

Though some areas of specialization are currently showing little to no job growth, retirement trends may help place new candidates.

Specialty	10 yr Job Growth	Median Salary	Mid-90-Percentile
Aerospace	0%	$61,530	$40,000 - $87,000
Chemical	9%	$42,920	$26,000 - $71,000
Civil	1%	$47,560	$30,000 - $72,000
Electrical/Electronics	0%	$57,850	$35,000 - $83,000
Electro-Mechanical	4%	$51,820	$33,000 - $77,000
Environmental	18%	$45,350	$29,000 - $77,000
Geological/Petroleum	15%	$52,700	$27,000 - $99,000
Industrial	-3%	$50,980	$33,000 - $76,000
Mechanical	5%	$51,980	$33,000 - $77,000
Nuclear	15%	$69,060	$42,000 – $97,000

Environmental, Safety, & Health Engineer

Description:

Environmental engineers assess current environmental damage, develop solutions to address existing problems, and evaluate future projects for impacts to water, wildlife, air, and people both locally and globally. Safety and health engineers focus more on predicting human health effects from proposed work. They assess the chemical, mechanical, and electrical safety hazards of a given program and develop ways to ensure human safety.

Sample Day-to-Day Tasks:

- Assess air and water pollution as well as hazardous waste at a site
- Develop a system to minimize or eliminate environmental damage
- Propose techniques to reuse, recycle, and safely discard of manufacturing by-products
- Assess projects for chemical, mechanical, and electrical safety for manufacturing staff and end users
- Develop regulations to promote safety and safety awareness
- Propose alternate methods of production when a potential hazard is identified
- Prepare reports and present to management and employees
- Perform inspections to ensure compliance

Work Environment:

Engineers typically work in offices with ~40 hour weeks. Sometimes engineers are required to travel to perform on-site installations and inspections. Jobs can be found in both military and civilian venues. Engineers can work in the research, development, or manufacturing areas though the latter tends to employ the most ESH engineers. Engineers can perform as technical contributors, program managers, or functional managers and supervisors.

Required Training:

- Entry level requires a B.S. in a related engineering discipline.
- A graduate of a four year technology program may qualify for entry level positions, as well, but cannot obtain a professional engineering (P.E.) license. P.E. certification is required to work directly with the public which may be required of ESH engineers.
- Continual learning through on the job training, journal articles, technical conferences, and company funded classes is expected throughout one's career.

Employment Prospects/Pay:

With increased emphasis on prevention of environmental and safety issues, the job growth for environmental engineers is expected to rise 15% over the next decade, and that for safety and health engineers will rise 11% in the same time frame.

The median salary in 2012:
- $80,890 for environmental engineers
- $76,830 for safety and health engineers

Mid-90-percentile range:
- $50,000 - $122,000 for environmental engineers
- $45,000 - $119,000 for safety and health engineers

The average 2012 salary by top industries for environmental:
- $98,890 in federal government
- $81,900 in engineering services
- $77,000 in consulting

The average 2012 salary by top industries for safety & health:
- $75,870 in professional & technical services
- $75,180 in state & local government
- $70,420 in construction

Industrial Engineer

Description:

Industrial engineers increase production efficiency through business organization, people management, manufacturing and information systems, planning, quality assurance, and cost analyses. As their jobs are very similar in nature to that of business management, they often move into supervisory and executive positions quickly.

Sample Day-to-Day Tasks:

- Use computer assisted design tools to design equipment and machinery layouts
- Determine best practices for manufacturing processes, material and process flow, and quality assurance
- Review and advise on staff utilization, design standards, and procurement
- Perform detailed calculations and apply statistical methods
- Coordinate and direct management and manufacturing
- Perform inspections to ensure compliance with safety and operation protocols
- Work with other engineers and customers to provide customized solutions
- Supervise and train personnel
- Develop cost and schedule estimates for program and process improvements

Work Environment:

Engineers typically work in offices with ~40 hour weeks. Sometimes engineers are required to travel to perform on-site installations or confer with customers. Desk work as well as hands on work is expected, though the mix varies by job and field. Jobs can be found in both military and civilian sectors. These engineers typically work in manufacturing areas but can be of great use in other areas, as well. Engineers can also perform as technical contributors, program managers, or functional managers and supervisors.

Required Training:

- Entry level requires a B.S. in a related engineering discipline.
- A graduate of a four year technology program may qualify for entry level positions, as well, but cannot obtain a professional engineering (P.E.) license. P.E. certification is required to work directly with the public but is not required for most industry jobs.
- Research and development professionals as well as college faculty in engineering are required to have an M.S. or usually a Ph.D.
- Continual learning through on the job training, journal articles, technical conferences, and company funded classes is expected throughout one's career.

Employment Prospects/Pay:

Job growth in this area is expected to increase only 5% over the next decade, about half the average of all other occupations. However, industrial engineers enjoy opportunities in a variety of other fields.

The median salary in 2012:
- $78,860

Mid-90-percentile range:
- $51,000 - $118,000

The average 2012 salary by top industries:
- $84,600 in aerospace manufacturing
- $82,290 in management
- $81,240 in engineering services

Starting salaries for new industrial engineering baccalaureates in 2009:
- $58,358

Marine Engineer or Naval Architect

Description:

Marine engineers design ship components and instrumentation including navigation and propulsion systems. Naval architects focus on the structural design of marine vessels to meet form and stability requirements. Both professions contribute to the development of submarines, recreational- and industrial-use ships, aircraft carriers, and tankers.

Sample Day-to-Day Tasks:

- Use computer assisted design tools to prepare technical drawings
- Perform detailed calculations of operating parameters and specifications for a given project
- Design structure, navigation controls, or instrumentation for sea-faring vessels
- Design and perform various experiments to test designs and equipment compliance
- Coordinate investigation of technical issues with design implementations
- Perform inspections to ensure compliance with safety and operation protocols
- Work with other engineers and customers to provide customized solutions
- Supervise and train personnel
- Develop cost and schedule forecasts for programs

Work Environment:

Engineers typically work in offices with ~40 hour weeks. Sometimes engineers are required to travel to perform on-site installations or confer with customers. Desk work as well as lab work is expected, though the mix varies by job title and personal preference. Jobs can be found in both military and civilian venues. Engineers can work in

the research, development, or manufacturing areas. Engineers can also perform as technical contributors, program managers, or functional managers and supervisors. Opportunities for publications and patents are high.

Required Training:

- Entry level requires a B.S. in a related engineering discipline.
- A graduate of a four year technology program may qualify for entry level positions, as well, but cannot obtain a professional engineering (P.E.) license. P.E. certification is required to work directly with the public but is not required for most industry jobs.
- Research and development professionals as well as college faculty in engineering are required to have an M.S. or usually a Ph.D.
- Continual learning through on the job training, journal articles, technical conferences, and company funded classes is expected throughout one's career.

Employment Prospects/Pay:

Job growth is expected to be 10% over the next decade, about the average for all occupations. Competition is expected to be low given the small number of people entering this field.

The median salary in 2012:
- $88,100

Mid-90-percentile range:
- $54,000 - $151,000

The average 2012 salary by top industries:
- $97,550 in federal government
- $92,010 in engineering services
- $89,220 in water transportation
- $82,510 in ship building

Materials Engineer

Description:

Materials engineering impacts just about every area of new technology. These engineers develop and test anything from semi-conductors and ceramics to metals and plastics. They tend to concentrate their studies toward a particular material whether it's semi-conductors for computer chips, alloys for recreational and exercise equipment, or ceramics for glass and optical fiber. Materials engineers tend to work in the R&D sector.

Sample Day-to-Day Tasks:

- Use computer assisted design tools to mimic atomic interactions
- Develop new materials that meet mechanical, electrical, and chemical specifications
- Prepare technical drawings
- Design processing methods and equipment as well as testing protocols
- Assess material performance and degradation
- Work with other engineers and customers from various industries to provide customized solutions
- Supervise and train personnel
- Develop cost and schedule forecasts for programs

Work Environment:

Engineers typically work in offices with ~40 hour weeks. Sometimes engineers are required to travel to perform on-site installations or confer with customers. Desk work as well as lab work is expected, though the mix varies by job title and personal preference. Jobs can be found in both military and civilian venues. Engineers can work in

the research, development, or manufacturing areas. Engineers can also perform as technical contributors, program managers, or functional managers and supervisors. Opportunities for publications and patents are high.

Required Training:

- Entry level requires a B.S. in a related engineering discipline.
- A graduate of a four year technology program may qualify for entry level positions, as well, but cannot obtain a professional engineering (P.E.) license. P.E. certification is required to work directly with the public but is not required for most industry jobs.
- Research and development professionals as well as college faculty in engineering are required to have an M.S. or usually a Ph.D.
- Continual learning through on the job training, journal articles, technical conferences, and company funded classes is expected throughout one's career.

Employment Prospects/Pay:

Despite an increase in biotechnology and nanotechnology applications, job growth in this area is not expected to increase over the next decade. However, retirement of an aging population should help employ new candidates.

The median salary in 2012:
- $85,150

Mid-90-percentile range:
- $53,000 - $130,000

Starting salaries for new materials engineering baccalaureates in 2009:
- $57,349

Mechanical Engineer

Description:

Mechanical engineering is the most diverse field in the engineering curricula. Concentrations within the major include thermodynamics, power systems, fluid mechanics, electro-magnetics, and static and dynamic mechanical systems. Mechanical engineers design and test anything from robots to manufacturing equipment to engines and so on. Because the field is so diverse, mechanical engineers often find work in many engineering disciplines as well as management.

Sample Day-to-Day Tasks:

- Use computer assisted design tools to prepare technical drawings
- Develop and test new mechanical structures
- Analyze designs for structural, thermal, and electrical performance
- Perform detailed calculations of operating parameters and specifications for a given project
- Test components for defects and provide fault analysis
- Coordinate and direct technicians or manufacturing staff in the implementation of designs
- Perform inspections to ensure compliance with safety and operation protocols
- Work with other engineers and customers to provide customized solutions
- Supervise and train personnel
- Develop cost and schedule forecasts for programs

Work Environment:

Engineers typically work in offices with ~40 hour weeks. Sometimes engineers are required to travel to perform on-site installations or confer with customers. Desk work as well as lab work is expected, though the mix varies by job title and personal preference. Jobs can be found in both military and civilian venues. Engineers can work in

the research, development, or manufacturing areas. Engineers can also perform as technical contributors, program managers, or functional managers and supervisors. Opportunities for publications and patents are high.

Required Training:

- Entry level requires a B.S. in a related engineering discipline.
- A graduate of a four year technology program may qualify for entry level positions, as well, but cannot obtain a professional engineering (P.E.) license. P.E. certification is required to work directly with the public but is not required for most industry jobs.
- Research and development professionals as well as college faculty in engineering are required to have an M.S. or usually a Ph.D.
- Continual learning through on the job training, journal articles, technical conferences, and company funded classes is expected throughout one's career.

Employment Prospects/Pay:

Consumer product demand and technology expansion will contribute to job growth of about 5% over the next decade. Though less than the average for all occupations, many mechanical engineering majors find work in related fields that carry a slightly different job title like aerospace, industrial, or similar. Also, there is an extreme lack of eligible engineers with U.S. citizenship that qualify for jobs with the government or with government contractors.

The median salary in 2012:
- $80,580

Mid-90-percentile range:
- $52,000 - $122,000

Starting salaries for new mechanical engineering baccalaureates in 2009:
- $58,766

Mining, Nuclear, or Petroleum Engineer

Description:

Mining, nuclear, and petroleum engineers design the facilities and processes associated with extracting and creating energy. Mining and geological engineers focus on coal, minerals, and metals (including non-energy related materials such as gold). Nuclear engineers focus on nuclear energy and radiation. And, petroleum engineers focus on oil and gas. All of these engineers must balance efficiency of extraction with human and environmental safety in their designs.

Sample Day-to-Day Tasks:

- Propose new means of detection of energy resources
- Develop extraction processes that meet efficiency and safety protocols
- Design facilities that operate in extraction of raw materials or in creation of energy
- Develop regulations to ensure minimal impact to workers, the environment, and the surrounding communities
- Assess new methods of energy distribution for more effective energy utilization
- Audit manufacturing processes for consistency in fabrication, safety, and government compliance
- Supervise and train personnel as needed
- Develop cost and schedule forecasts for programs

Work Environment:

Engineers typically work in offices with ~40 hour weeks. Sometimes engineers are required to travel to perform on-site installations or confer with customers. Desk work as well as lab work is expected, though the mix varies by job title and personal preference. Jobs can be found in both military and civilian venues. Engineers can work in

the research, development, or manufacturing areas. Engineers can also perform as technical contributors, program managers, or functional managers and supervisors. Opportunities for publications and patents are high.

Required Training:

- Entry level requires a B.S. in a related engineering discipline.
- A graduate of a four year technology program may qualify for entry level positions, as well, but cannot obtain a professional engineering (P.E.) license. P.E. certification is required to work directly with the public but is not required for most industry jobs.
- Research and development professionals as well as college faculty in engineering are required to have an M.S. or usually a Ph.D.
- Continual learning through on the job training, journal articles, technical conferences, and company funded classes is expected throughout one's career.

Employment Prospects/Pay:

The job growth for mining engineers is expected to be 12% over the next decade; that for nuclear engineers is expected to be 9%; that for petroleum engineers is expected to be 26%. In all cases, job competition is expected to be low.

The median salary in 2012:
- $84,320 (mining), $104,270 (nuclear), $130,280 (petroleum)

Mid-90-percentile range:
- $50,000 - $140,000 in mining engineering
- $69,000 - $150,000 in nuclear engineering
- $75,000 - >$187,000 in petroleum engineering

Starting salaries for new engineering baccalaureates in 2009:
- $64,404 (mining), $61,610 (nuclear), $83,121 (petroleum)

Surveyor

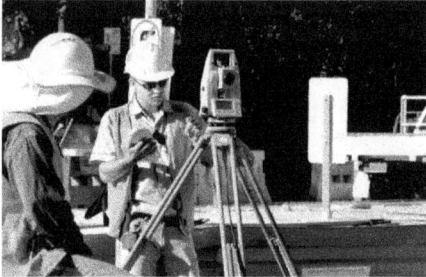

Description:

A surveyor uses various instruments to collect data about the Earth's surface features and property boundaries. They assist in mapping efforts, home building/planning, industrial construction, and engineering efforts.

Sample Day-to-Day Tasks:

- Use computer assisted design tools such as GPS to collect data
- Perform measurements of distance and height
- Interface with high tech equipment
- Perform calculations
- Interpret data and document results
- Mark legal property boundaries
- Research documentation about a specific site
- Keep detailed records
- Present findings in written and oral form to employers and government agencies
- Provide expert court testimony as needed

Work Environment:

Surveyors typically split their work between the office and the field with ~40 hour weeks. Long periods of standing and inclement weather are sometimes part of the job. Work is typically found in the civilian sector.

Required Training:

- Entry level requires a B.S. in surveying, civil engineering, forestry, or a related discipline.
- Licensing is required to work in public service and to certify legal documents.
- Licensing requirements vary by state but usually involve passing the Fundamentals of Surveying exam, working for 2 years with a licensed surveyor, then passing the Principles and Practice of Surveying exam.

Employment Prospects/Pay:

Job growth is expected to be 10% over the next decade, about the same as for all occupations.

The median salary in 2012:
- $56,230

Mid-90-percentile range:
- $32,000 - $91,000

The average 2012 salary by top industries:
- $68,590 in state government
- $61,880 in local government
- $57,250 in heavy construction
- $55,260 in mining, oil, & gas
- $54,430 in architectural services

Part 4: Mathematics

Though all of the careers discussed thus far require varying degrees of mathematics, a concentration in the field opens many opportunities. Examples of Mathematics careers include accountant, actuary, statistician, and financial manager.

Accountant or Auditor

Description:

Accountants and auditors are always in demand in a capitalistic society. Accountants can specialize in tax code, finance, or employee benefits to name a few. They can work for the public, for industry, or for the government. Auditors ensure records are accurate and regulations are observed. They can work internal or external to an organization they examine.

Sample Day-to-Day Tasks:

- Prepare and analyze accounting records and financial reports
- Use computer assisted tools to maintain records and analyze data
- Compute and file tax returns
- Advise clients in finance/investment strategies, compensation policies, and employee benefits
- Conduct internal or external reviews of financial records
- Summarize financial data and present to management
- Evaluate accounting processes for compliance and efficiency
- Analyze data trends and make recommendations to improve financial performance
- Supervise and train personnel

Work Environment:

Accountants and auditors work in offices with ~40 hour weeks, but tax accountants work very long hours during tax season depending on their number of clients. Some jobs require travel or offer telecommuting options. Jobs can be found in both military and civilian venues. Accountants and auditors are needed in any industry that turns a dollar. They can work as business owners, contractors, employees, program managers, and functional managers.

Required Training:

- Entry level positions require a B.S. in accounting or mathematics.
- Some companies prefer an M.S. or M.B.A., as well.
- Licensure as a certified public accountant (CPA) is required for many advanced positions.
- To become a CPA, four sections of the exam must be passed within 18 months. Less than half of applicants succeed on their first try.
- License renewal is required and contingent on a certain number of hours of continuing education.

Employment Prospects/Pay:

Business growth, changing regulations, and tighter auditing contribute to an expected job growth of 13% over the next decade. CPAs are expected to have the best employment potential. People with degrees in math related subjects usually have opportunities in a variety of fields. Going into business for yourself can far surpass the salary numbers reported below.

The median salary in 2012:
- $63,550

Mid-90-percentile range:
- $40,000 - $112,000

The average 2012 salary by top industries:
- $66,530 in finance & insurance
- $65,300 in manufacturing
- $64,670 in management
- $63,910 in accounting services
- $61,490 in government

Starting salaries for new math B.S. and M.S. graduates in 2009:
- $49,000

Actuary

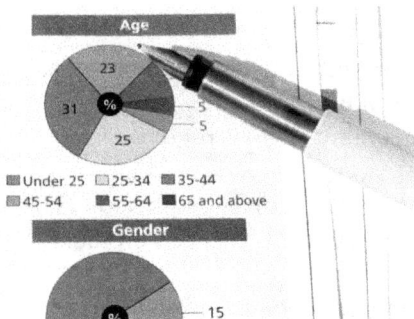

Description:

Actuaries assess the risk of potential outcomes to certain events. Areas of expertise include mathematics, statistics, finance, and business. Actuaries are instrumental to the insurance and financial planning industries to help minimize costs and maximize returns.

Sample Day-to-Day Tasks:

- Process and analyze data for modeling and analysis
- Discover data trends and summarize relationships
- Identify factors affecting reliability of data
- Summarize findings in written and presentation format
- Identify areas for improvement and growth
- Supervise, train, and hire personnel
- Develop cost and schedule forecasts for policy or program adjustments

Work Environment:

Actuaries work in offices with ~40 hour weeks. Some travel and extra hours may be required of consultants. Jobs are typically found in civilian venues particularly in the insurance and finance industries. Management opportunities are available.

Required Training:

- Entry level positions require at least a B.S. in mathematics, finance, statistics, or business but always require a solid math background (including calculus).
- Internships are recommended.
- In addition, full professional status is only obtained after several examinations are passed, usually taking 4-8 years to complete.
- Employers will sometimes hire a candidate fresh from school and help them with examination fees.
- Continual learning through on the job training, journal articles, technical conferences, and company funded classes is expected throughout one's career.

Employment Prospects/Pay:

Though jobs are anticipated to grow 26% over the next decade, qualified candidates are expected to meet with stiff competition. Consulting services will provide the most job growth. People with degrees in math related subjects usually have opportunities in a variety of fields.

The median salary in 2012:
- $93,680

Mid-90-percentile range:
- $56,000 - $175,000

Top industries:
- Insurance
- Professional services
- Management
- Financial investment
- Government

Budget Analyst

Description:

Budget analysts are concerned with the efficiency and profitability of a company's operations. In addition to overseeing budgets, they evaluate programs and sometimes draft budget legislation or testify before a fund-granting authority. Almost half of all budget analysts work for the government.

Sample Day-to-Day Tasks:

- Interpret and analyze budget and accounting reports
- Provide insight, recommendations, and assistance with cost analysis and budget preparation
- Analyze accounting data to estimate resource requirements
- Conduct cost-benefit analyses
- Make recommendations on funding requests
- Improve efficiency and profits

Work Environment:

Budget analysts work in offices with ~40 hour weeks. Some jobs require travel or offer telecommuting options. Jobs can be found in both military and civilian venues. Many analysts work for government, private industry, or non-profit organizations. Budget analysts can work as technical contributors, program managers, and functional managers.

Required Training:

- Entry level positions require a B.S. typically in accounting, finance, business, economics, or statistics.
- An M.S. is preferred and sometimes required.
- Continual learning through on the job training, journal articles, technical conferences, and company funded classes is expected throughout one's career.

Employment Prospects/Pay:

Efficient financial operations continue to be a focus in industry. Jobs for budget analysts are expected to increase 6% over the next decade, which is about half the average. People with degrees in math related subjects usually have opportunities in a variety of fields.

The median salary in 2012:
- $69,280

Mid-90-percentile range:
- $46,000 - $104,000

Top industries:
- Federal government
- Education services
- State government
- Manufacturing
- Professional services

Economist

Description:

Economists analyze how various resources are utilized in a society. Their study of land, labor, and materials enables them to predict economic trends, exchange rates, employment levels, and more. Economists can be divided into two main categories: microeconomists who focus on individuals or businesses and macroeconomists who look at historical and global economies.

Sample Day-to-Day Tasks:

- Analyze and interpret data in an area of specialization such as land, labor, materials, or finance
- Apply mathematical and statistical techniques to forecast markets
- Testify in legislative hearings on the impact of new policy
- Develop new analysis tools
- Use computer aided analysis to compile reports and make predictions
- Teach and supervise as needed

Work Environment:

Economists work in offices with ~40 hour weeks. Tight deadlines can require overtime work. Some jobs require travel or offer telecommuting options. Jobs can be found in the civilian sector in government and private industry. Management opportunities are available.

Required Training:

- Entry level positions usually require an M.S. or Ph.D. in economics.
- Some some entry level positions are available for those with a B.S.
- Advancement usually favors the graduate degree.
- Continual learning through on the job training, journal articles, technical conferences, and company funded classes is expected throughout one's career.

Employment Prospects/Pay:

Jobs for economists are expected to increase 14% over the next decade, faster than the average for all occupations. People with degrees in economics are successful at finding jobs in related fields.

The median salary in 2012:
- $91,860

Mid-90-percentile range:
- $51,000 - $155,000

The average 2012 salary by top industries:
- $110,580 in finance & insurance
- $106,850 in federal government
- $94,630 in research & development
- $91,570 in consulting
- $63,880 in state & local government

Cost Estimator

Description:

Cost estimators are responsible for estimating the money required to complete large projects. They factor in costs of materials, equipment, labor, and other resources as well as the time to finish the project as pertains to rents, leases, permits, and depreciation. Two areas of specialization include construction and manufacturing.

Sample Day-to-Day Tasks:

- Research historical data to draw conclusions about current project
- Use advanced software tools to access databases and develop simulations
- Travel to sites and interact with project managers to assess scope of effort
- Develop project plans
- Propose cost saving measures
- Document estimates and analyses and present to management
- Coordinate effort with a staff of estimators on large projects

Work Environment:

Cost estimators work in offices and in the field with ~40 hour weeks and some local travel. High pressure demands and overtime work are a typical aspect of the job for most industry employees. Jobs are usually found in the civilian sector in construction and manufacturing. Cost estimators can work as independent contractors or employees of larger firms. Management opportunities are available.

Required Training:

- Entry level positions usually require a B.S. in a math intensive field as well as related industry experience. Sometimes enough experience can waive the education requirement.
- Experience credentials are earned in related industry work, internships, and/or on the job training.
- Certification is required by some employers.
- Certification involves passing an exam after attaining 2 years experience and may require publication of a technical article.
- Continual learning through on the job training, journal articles, technical conferences, and company funded classes is expected throughout one's career.

Employment Prospects/Pay:

Job growth is expected to increase by 26% over the next decade, over double that for the average of all occupations. Construction and infrastructure improvements are expected to create the largest demand. People with degrees in math related subjects usually have opportunities in a variety of fields.

The median salary in 2012:
- $58,860

Mid-90-percentile range:
- $35,000 - $97,000

Financial Analyst

Description:

Financial analysts are also referred to as securities analysts and investment analysts. They guide businesses and clients through the maze of stocks, bonds, commodities, and other prospects. They analyze current trends, forecast financial performance, and develop investment strategies.

Sample Day-to-Day Tasks:

- Analyze financial performance data
- Advise on investment strategies
- Monitor and forecast developments in various business/technology sectors
- Prepare reports and presentations
- Determine securities prices
- Compare investment vehicles for performance and risk
- Supervise and train personnel

Work Environment:

Financial analysts work in offices with ~40 hour weeks, but many work long hours and travel frequently. High pressure demands are a typical aspect of the job for most industry employees. Jobs are usually found in the civilian sector. Financial analysts can work as business owners, contractors, employees, and functional managers.

Required Training:

- Entry level positions require a B.S. in finance, business, accounting, or similar.
- Many companies prefer an M.S. or M.B.A. with a strong math background, especially for advancement.
- Licensure and certifications are sometimes required for advancement. Such additional requirements are pursued after the candidate has obtained some job experience.
- Continual education is a necessary part of the career path especially when evaluating other business and technology sectors for investment viability.

Employment Prospects/Pay:

Job growth is expected to increase by 16% over the next decade, faster than that for the average of all occupations. Competition is expected to be high especially for candidates with no job experience.

Performance bonuses commonly contribute significant additional amounts to the base wages shown below.

The median salary in 2012:
- $76,950

Mid-90-percentile range:
- $47,000 - $148,000

The average 2012 salary by top industries:
- $90,560 in financial investment services
- $75,920 in professional services
- $75,300 in credit services
- $75,200 in management
- $72,270 in insurance services

Financial Examiner

Description:

Financial examiners oversee a bank's financial operations to ensure compliance with regulation and proper management. They assess the bank's financial position and assess loans in terms of risk to the bank and fairness to the consumer. They make recommendations on future decisions that impact the strength of a financial institution.

Sample Day-to-Day Tasks:

- Analyze and interpret balance sheets and other documentation
- Assess the financial strength of an institution in terms of allowable risk, cash on hand, and other metrics
- Ensure fair lending practices
- Safeguard against overextending on risk
- Propose methods to deal with uncertain financial conditions
- Ensure regulatory compliance
- Assess new regulations for effects on current financial health
- Train and supervise personnel as needed

Work Environment:

Financial examiners work in offices with ~40 hour weeks with some local travel required. Jobs can be found in the civilian sector in government, finance, banking, and insurance. Management opportunities are available.

Required Training:

- Entry level positions require a B.S. with coursework in finance and accounting.
- On the job training is provided for junior employees.
- Advancement opportunities usually require an M.S. in accounting, an M.B.A., or becoming a certified public accountant (CPA).
- Continual learning through technical conferences, regulation literature, and company funded classes is expected throughout one's career.

Employment Prospects/Pay:

Job growth is expected to increase 6% over the next decade, slower than the average for all occupations. People with degrees in math related subjects usually have opportunities in a variety of fields.

The median salary in 2012:
- $75,800

Mid-90-percentile range:
- $43,000 - $141,000

Top industries:
- Federal government
- Credit services
- State government
- Investment services

Financial Manager

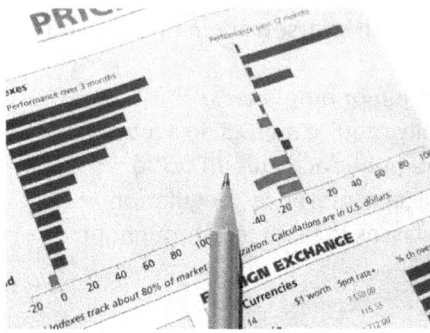

Description:

Financial managers oversee accounting reports, investments, and cash management. Specialties include controller (report generation), treasurer (budget direction), and credit, cash, or risk manager (assessment and direction). Financial managers are needed in finance, insurance, banking, and government.

Sample Day-to-Day Tasks:

- Analyze and interpret data
- Oversee financial report generation and ensure accuracy
- Oversee investments, cash flow, and budgets
- Create policy to ensure good fiscal health
- Manage departments and personnel to meet financial goals
- Inform as to recommended path to efficient, profitable operation

Work Environment:

Financial managers work in offices with ~40 hour weeks, though overtime is usually required. Some jobs require travel or offer telecommuting options. Jobs can be found in the civilian sector in government, insurance, finance, banking, and other private industry.

Required Training:

- Entry level positions require a B.S. in finance, accounting, economics, or business administration.
- Emphasis is placed on experience, and higher degrees are sometimes preferred.
- Continual learning through on the job training, journal articles, technical conferences, and company funded classes is expected throughout one's career.

Employment Prospects/Pay:

Jobs for financial managers are expected to increase 9% over the next decade with competition for jobs likely. People with degrees in math related subjects usually have opportunities in a variety of fields. The wages listed below do not include stock options and cash bonuses.

The median salary in 2012:
- $109,740

Mid-90-percentile range:
- $60,000 - >$187,000

The average 2012 salary by top industries:
- $130,120 in professional services
- $124,840 in management
- $108,690 in finance & insurance
- $107,730 in manufacturing
- $102,270 in government

Insurance Underwriter

Description:

Insurance underwriters determine what kinds of insurance a company will grant and under what terms. They set premiums, rates, and range of coverage. They approve or disapprove applications with a careful balance of risk and conservatism. Specialties include life insurance, health insurance, mortgage insurance, and property and casualty insurance.

Sample Day-to-Day Tasks:

- Review and analyze insurance applications
- Verify information submitted by a potential client
- Confer with other experts as needed
- Use computer assisted tools to aid in decision making
- Decide to accept or deny an applicant
- Write the insurance policy setting rates, premiums, and coverage
- Balance decisions against income producing risk and cautious cash protection

Work Environment:

Insurance underwriters typically work in offices with ~40 hour weeks, though some local travel may be required to view assets in person. Jobs can be found in the civilian sector in industries which provide insurance. Management opportunities are available.

Required Training:

- Entry level positions require a B.S. in finance, accounting, economics, or business administration.
- On the job training is provided for junior employees.
- Certification is usually required for advancement and may take up to 2 years of additional coursework.
- Continual learning through on the job training, technical conferences, and company funded classes is expected throughout one's career.

Employment Prospects/Pay:

Job growth is expected to increase 6% over the next decade, which is about half the average for all occupations. Automation and new software tools are reducing the need for underwriters, but qualified people will still be needed to check computer generated recommendations. People with degrees in math related subjects usually have opportunities in a variety of fields.

The median salary in 2012:
- $62,870

Mid-90-percentile range:
- $39,000 - $110,000

Market Research Analyst

Description:

Market research analysts help companies target product/service development and advertising more efficiently. They gain insight into various market sectors, geographic or demographic, and evaluate them for fitness to a company's offering. Surveys, interviews, focus groups, questionnaires, and analysis of competitors are typical tools of the trade.

Sample Day-to-Day Tasks:

- Conduct research into buying habits and preferences of a market sector
- Develop and implement questionnaires, surveys, focus groups, and interviews
- Assess market position of competition
- Use computer assisted tools to perform statistical analyses on acquired data
- Forecast trends and successful marketing strategies
- Make recommendations to management

Work Environment:

Market research analysts typically work in offices with ~40 hour weeks, though overtime is sometimes needed to meet tight deadlines. Direct interaction with the public is usually involved. Jobs can be found in the civilian sector, mostly in private industry. Direct corporate employment or working through consulting services comprise most of the occupations. About 5% are self-employed. Management opportunities are available.

Required Training:

- Entry level positions require a B.S. in finance, accounting, economics, or business administration.
- Advancement usually requires an M.S. or M.B.A.
- A Ph.D. is required for university faculty positions.
- Certification can be an asset and requires 3 years experience, passing an exam, and certification renewal every 2 years.
- Continual learning through on the job training, journal articles, technical conferences, and company funded classes is expected throughout one's career.

Employment Prospects/Pay:

With companies seeking to cut costs while still raising revenues, job growth is expected to increase 32% over the next decade, which is about triple the average for all occupations. People with degrees in math related subjects usually have opportunities in a variety of fields.

The median salary in 2012:
- $60,300

Mid-90-percentile range:
- $33,000 - $114,000

The average 2012 salary by top industries:
- $67,550 in manufacturing
- $67,330 in management
- $64,490 in finance & insurance
- $60,200 in wholesale trade
- $56,760 in consulting services

Mathematician

Description:

One of the oldest STEM professions, a mathematician enjoys a wide range of career opportunities. Theoretical mathematicians discover new principles and relationships in the field and are usually researchers and professors at universities. Applied mathematicians use theories and computational techniques to solve a variety of problems in the engineering and physical sciences.

Sample Day-to-Day Tasks:

- Discover new theoretical principles and relationships in mathematics
- Publish discoveries and share findings with other professionals
- Apply theories and computational techniques to real world problems
- Perform complex computations and numerical analysis techniques
- Develop and analyze encryption systems
- Conduct research and teach college level classes
- Manage and advise graduate students

Work Environment:

Mathematicians work in offices with ~40 hour weeks. Jobs can be found in both military and civilian venues. Work is usually in the research and development areas for industry or academia. Mathematicians can work as technical contributors, program managers, functional managers, and professors. Opportunities for publications and patents are high.

Required Training:

- Entry level positions require a Ph.D. in mathematics for private industry and academia.
- Some federal government positions only require a B.S. in mathematics.
- Additional degrees or training improve chances of work in related fields such as engineering and the physical sciences.
- Continual learning through on the job training, journal articles, technical conferences, and classes is expected throughout one's career.

Employment Prospects/Pay:

Though jobs are anticipated to grow 23% over the next decade, qualified candidates are expected to meet with significant competition. However, people with degrees in mathematics find work in a variety of fields.

The median salary in 2012:
- $101,360

Mid-90-percentile range:
- $56,000 - $153,000

The average 2012 salary by top industries:
- $118,030 in research & development
- $116,860 in manufacturing
- $106,360 in federal government
- $74,980 in management
- $66,590 in education services

Operations Research Analyst

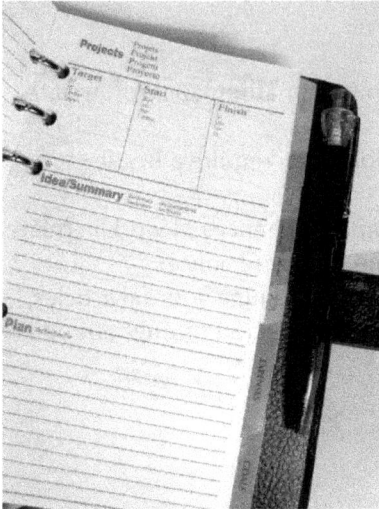

Description:

Operations research analysts study the efficiency of systems and procedures to meet specified goals. They analyze data to make recommendations on policy and improve performance. Operations research analysts solve problems related to finance, people, materials, and time. They have contributed to military operations as well as private business interests. A strong analytical skillset is needed for this position.

Sample Day-to-Day Tasks:

- Develop mathematical models and simulations
- Validate and test models for accuracy
- Coordinate with management to identify and resolve issues
- Develop test protocols and data requirements
- Prepare reports and give presentations on recommended courses of action
- Develop business systems for improved operations

Work Environment:

Operations research analysts work in offices with ~40 hour weeks. Tight deadlines can sometimes require overtime work. Some jobs require travel or offer telecommuting options. Jobs can be found in the civilian and military sectors. Management opportunities are available.

Required Training:

- Some entry level positions are available to persons with a B.S., but an M.S is usually preferred.
- Coursework should emphasize advanced mathematics and computer skills.
- Continual learning through on the job training, journal articles, technical conferences, and company funded classes is expected throughout one's career.

Employment Prospects/Pay:

Jobs for operations research analysts are expected to increase 27% over the next decade, more than double the average of all other occupations. People with degrees in math related subjects also enjoy opportunities in a variety of fields.

The median salary in 2012:
- $72,100

Mid-90-percentile range:
- $41,000 - $129,000

The average 2012 salary by top industries:
- $79,630 in manufacturing
- $74,490 in computer systems design services
- $72,630 in management
- $67,480 in finance & insurance
- $56,670 in state & local government

Personal Financial Advisor

Description:

Personal financial advisors help individuals with decisions about savings, investment, insurance, and taxes. They offer advice or manage people's finances for short term goals like college expenses or long term goals like retirement. They work with clients from a range of financial backgrounds.

Sample Day-to-Day Tasks:

- Advertise services and network to find clients
- Assess a client's financial standing and goals
- Develop plans and strategies for meeting financial objectives
- Educate clients on their options
- Manage client investments in stocks, bonds, or other opportunities
- Research economic trends, market trends, and insurance offerings
- Advocate insurance as needed
- Determine tax implications for various strategies
- Monitor finances and offer new direction when necessary

Work Environment:

Personal financial advisors work in offices with ~40 hour weeks, though about a quarter work more than 50 hours. Advertising services, networking, and conference attendance is important to building clientele and may require evening or weekend work. Direct interaction with clients is a key part of the job. Work can be found in the civilian sector with about one quarter self-employed.

Required Training:

- Entry level positions require a B.S. in finance, accounting, economics, law, or business administration.
- Advancement usually requires an M.S. or M.B.A.
- Licensing and state or SEC registration is required to buy or sell stocks, bonds, and insurance policies.
- Certification can be an asset and requires 3 years experience and passing an exam.
- Continual learning through on the job training, journal articles, technical conferences, and company funded classes is expected throughout one's career.

Employment Prospects/Pay:

With an aging population and more soon-to-be retirees, jobs for personal financial advisors are expected to increase 27% over the next decade, almost triple the average of all other occupations. People with degrees in math related subjects usually have opportunities in a variety of fields

Bonuses and self-employed business owners are not represented in the wage numbers below.

The median salary in 2012:
- $67,520

Mid-90-percentile range:
- $32,000 - >$187,000

The average 2012 salary by top industries:
- $83,400 in investment services
- $82,360 in professional services
- $72,630 in securities & commodities
- $63,500 in insurance
- $47,780 in credit services

Professor

Description:

Professors can be found in any STEM discipline. College and university faculty instruct students in their field of expertise. They also many times conduct research and serve on graduate committees. Assistant professors spend a significant amount of time on instruction whereas full professors, who have met the rigors of obtaining tenure, concentrate more on research opportunities.

Sample Day-to-Day Tasks:

- Develop lesson plans, grade papers, give assignments
- Lecture in small or large groups
- Write proposals for grant money for research
- Conduct research in a lab environment
- Direct graduate assistants in the completion of their programs
- Supervise graduate assistants in carrying out research programs
- Publish research in peer reviewed journals

Work Environment:

Professors work at colleges and universities with ~40 hour weeks. Tight deadlines for research or grading of final exams can sometimes require overtime work. Summer work involves lighter teaching loads, but research is usually on-going. Jobs are found in the civilian sector. Management opportunities are available. Opportunities for publications and patents are high.

Required Training:

- A Ph.D. with at least 18 graduate credit hours in the field of expertise is required to become an assistant professor.
- Getting tenure is a rigorous process which includes excellent job performance, research work, and publications.
- The tenure process can take 7 years to complete while working as an assistant or associate professor.
- Continual learning through on the job training, journal articles, technical conferences, and funded classes is expected throughout one's career.

Employment Prospects/Pay:

Jobs for post-secondary teachers are expected to increase approximately 19% over the next decade, almost double the average of all other occupations. Salary varies greatly with field of expertise, geographical location, and type of institution. Full professors generally make twice as much as assistant professors.

The median salary for all post-secondary teachers in 2012:
- $68,970

Mid-90-percentile range:
- $36,000 - $142,000

The average 2012 salary by specialty (abbreviated list):
- $92,670 in engineering
- $87,950 in economics
- ~$82,000 in earth & space related sciences
- ~$81,000 in agricultural and health sciences
- $78,540 in physics
- $77,320 in environmental sciences
- $74,180 in biology
- $72,200 in computer technology
- ~$71,000 in chemistry and architecture
- $64,990 in mathematics

Purchasing Manager

Description:

A purchasing manager, buyer, or purchasing agent ensures supply of merchandise to companies for use or resale. They negotiate contracts and check for quality. They evaluate suppliers for reputation, speed, and cost.   They identify new or back up suppliers for critical items and ensure the supply channel is always operational.

Sample Day-to-Day Tasks:

- Identify and evaluate potential suppliers
- Visit plants to inspect and learn about product manufacturing
- Network with colleagues and suppliers to extend contacts
- Negotiate contracts
- Ensure both parties are meeting the terms of the contract
- Inspect goods for quality and compliance
- Evaluate financial statements and supply reports
- Meet with suppliers to discuss defects and other issues
- Keep detailed records

Work Environment:

Purchasing managers work in offices with ~40 hour weeks. Overtime is usually expected, however. Some jobs require travel across the country or abroad. Stress can be high when suppliers fail to meet contractual deadlines or quality of delivered products is unacceptable. Work is usually found in the civilian sector.

Required Training:

- Entry level positions in some companies can be acquired with a high school education and on the job training.
- Most large corporations require a B.S. in business or accounting. For some employers, favor is given to those who major in engineering or the applied sciences.
- Continual learning through on the job training, technical conferences, and company funded classes is expected throughout one's career.

Employment Prospects/Pay:

Job growth is expected to be 4% over the next decade, slower than the average for all occupations. People with degrees in math or applied sciences usually have opportunities in a variety of fields.

The median salary in 2012:
- $60,550

Mid-90-percentile range:
- $35,000 - $110,000

The average 2012 salary by position:
- $100,170 for purchasing managers
- $58,760 for general purchasing agents
- $55,720 for purchasing agents of farm products
- $51,470 for purchasing agents in retail & wholesale

Statistician

Description:

Statisticians compile and organize vast amounts of numerical data into meaningful results. They are needed in many industries including engineering, physical sciences, marketing, and manufacturing. From designing research experiments to interpreting data, anything performed on a large scale needs a statistician.

Sample Day-to-Day Tasks:

- Interpret and analyze vast amounts of numerical data
- Discover data trends and summarize relationships that affect real world problems
- Identify factors affecting reliability of data
- Summarize findings in written and presentation format
- Design research experiments or quality control measures
- Define sampling techniques and identify candidate sample groupings
- Supervise and train personnel

Work Environment:

Statisticians work in offices with ~40 hour weeks. Some jobs require travel or offer telecommuting options. Jobs can be found in both military and civilian venues. Many statisticians work for government, insurance, research, or manufacturing facilities. Statisticians can work as technical contributors, program managers, and functional managers. Opportunities for publications in the research area are high.

Required Training:

- Entry level positions require a M.S. in mathematics or statistics.
- A Ph.D. is required for research and academia.
- Continual learning through on the job training, journal articles, technical conferences, and company funded classes is expected throughout one's career.

Employment Prospects/Pay:

With population growth, decreased margins required in business, and more thorough research analysis needs, jobs for statisticians are expected to increase by 27% over the next decade. People with degrees in math or statistics usually have opportunities in a variety of fields.

The median salary in 2012:
- $75,560

Mid-90-percentile range:
- $42,000 - $122,000

The average 2012 salary by top industries:
- $97,250 in federal government
- $69,850 in finance & insurance
- $66,210 in education services
- $63,420 in health care
- $50,860 in state & local government

Appendices

A: Picking a Career
B: The Cost of College

Appendix A: Picking a Career

Picking a profession for the rest of your life may seem like a daunting task. It is. However, when you make good money, you get to have interesting mid-career options. You may decide to go into management, specialize in an area slightly different than where you started, change gears entirely and go back to school, or perhaps even take a sabbatical to write books on the beach.

So, take a deep breath and ask yourself some questions. There are no right or wrong answers.

* Do you like to be hands on or work at a desk?
* Do you need to be creative?
* Are you drawn to natural sciences, medical sciences, or technology?
* Do you prefer a profession that is heavy or light in math?
* Do you prefer a profession that is heavy or light in memorization?
* Do you like to travel or even move to new locations?
* Do you want a career that is employable anywhere you go?
* Do you want to live in a city or in a rural location?
* Are you open to working abroad?
* Do you want to be able to telecommute (work from home)?
* Do you thrive on or buckle under stress and deadlines?
* Do you want to make life/death decisions?
* Do you like to be the leader or the follower?
* Do you like to work alone or in teams?
* Do you like to work with strangers (like customers or patients)?
* Do you like to work with animals?
* Do you want to have a direct impact on people, animals, or the planet?
* Do you want to make non-essentials like toys, games, and gadgets?
* Do you like to learn?
* Do you like responsibility and ownership?
* Do you like to explore new ideas or make current ones more efficient?
* Do you like the bragging rights of a specific job title?
* Do you want to work for yourself or for a company?
* How important are salary and benefits?
* Do you want to get started asap, or is going to school for awhile okay?

After answering these questions, pick the ones that matter most to you and measure your prospective careers against those criteria. Be flexible, but be honest with yourself.

Appendix B: The Cost of College

You may have seen reports or know people who say that college is a scam that cripples their graduates with debt. This is not true of all colleges and all majors. Yes, people who major in a field without realizing that they need a Master's or Doctorate to find a decent paying job will graduate with high debt and an unmarketable degree. Also, people who go to an outlandishly priced school may find the increased debt outweighs whatever increased starting salary their expensive school promised. Keep in mind, too, that by and large, your salary mid-career is dictated more by your performance and experience than by the name of the school you attended in your early 20's.

The bottom line is this: College is nothing more than a vocational school like Lincoln Tech or ITT; it's main purpose is to train you for a job!

With that in mind, how do you pick a college or university? First, make sure it is accredited and offers your selected major. It's a good idea to see if they have many fields of study as most freshmen change their majors. Second, ask the school about their job placement statistics in your field of study. They should be bragging about those statistics and have them at the ready! Third, if you are paying for it, make sure you can afford it in the long haul.

To answer whether you can afford the school, let's first look at how students pay for college. The two main avenues are military and civilian. There are three military options (see http://www.myfuture.com/military/articles-advice/college-assistance):

- Before service
    - Reserve Officers' Training Program (ROTC) (see http://www.military.com/rotc)
    - Service academies or military colleges
- During service
    - Coursework while enlisted in the military
- After service
    - GI bill and other funds

All of the above options provide the opportunity to receive full tuition, room and board, medical, and book expenses. The before service options provide regular college instruction as well as leadership courses that train the student to be an officer. If scholarships are accepted, the student is required to complete 4-5 years of service upon graduation. The ROTC program is available at over 1100 colleges across the U.S., or the student can attend a service academy or military college (like the U.S. Naval Academy). Programs are available in the Army, Air Force, and Naval branches of the military.

There are many programs for enlisted personnel that allow college study during service. Assistance is based on years of active duty and other eligibility requirements. As an enlisted person, room and board are already covered along with a salary.

For those who have already served in active duty in the last 15 years, the GI bill provides financial assistance for college. Each military branch also supports college fund programs for additional help.

The military route scares some because of the commitment to service, but it does ensure you have a job when you graduate and can eliminate having to start your adult life burdened with debt. Know, too, that students can enroll in part of the "before service" options for the leadership training without accepting scholarship money to avoid service requirements. It may then be possible to secure an officer position upon graduation.

There are also three civilian options:

- Scholarships for special abilities, interests, or achievements
    - Check out http://www.collegescholarships.org/ and http://www.studentscholarships.org/
- Need-based grants and loans
    - Fill out the Free Application for Federal Student Aid (FAFSA) which you can find at https://fafsa.ed.gov/
- Stafford loans (also must fill out the FAFSA)

The first two options are competitive, and many students don't qualify. Most students end up going the third route: Stafford loans. Stafford loans are granted to U.S. citizens, permanent residents, and eligible non-citizens who are enrolled in college at least half time. Unless you are in default (are not making expected payments) on a current school loan, credit is not a factor on receiving the loan, and you don't need a cosigner (a credit-worthy person willing to pay the loan if you do not).

There are two types of Stafford loans: Subsidized and Unsubsidized. Both types of loans are for tuition, housing, food, books, and required expenses. Both types of loans do not require you to pay anything until 6 months after you are no longer enrolled at least half time. You can also request additional deferments (excused periods of non-payment) upon certain conditions.

- For subsidized loans, the federal government pays the interest on the loan while you are in school. When you graduate, you owe what you borrowed and start paying the interest yourself.
- For unsubsidized loans, you have the choice of paying interest while in school, or capitalizing the interest until you graduate. Capitalizing means that the interest is continually added to your original loan amount (principle) until you start paying. When you graduate, you owe more than you borrowed if you have not paid the interest.

There are limits to how much you can borrow based on your year in school and whether or not you are independent. Gaining independent status unfortunately has nothing do with whether you are financially independent. You are considered independent if you meet certain criteria outlined at https://studentaid.ed.gov/fafsa/filling-out/dependency. For most students, the major criteria are:

- You are over 24 years old,
- You are a court emancipated minor, or
- You are married or supporting children, or
- You were under court ordered legal guardianship or were a ward of the state (not incarcerated) after you turned 13 years old, or
- Your parents are incarcerated, missing, or other.

In general, if you are not considered independent, your parents' financial information is taken into account on the FAFSA to determine if you qualify for need-based grants and loans. You also must be independent to get a subsidized Stafford loan.

For many, the unsubsidized Stafford loan is the only way to go (you still need to fill out the FAFSA every year you take out a new loan). The limits of borrowing in a given year for any Stafford loan are listed on the next page. If you qualify for subsidized loans, a portion of the loan limit will be eligible, but not all. The difference is then made up in unsubsidized loans.

Year in School	1	2	3+	Total Allowed
Dependent	$5,500	$6,500	$7,500	$31,000
Independent	$9,500	$10,500	$12,500	$57,500

If graduate school is in your future, aim for the best school you can get into that will still pay your tuition if that is available in your field of study. If you need to borrow, the following table gives the annual and total limits:

	Annual	Total Including Undergraduate
Graduate School	$20,500	$138,500
Medical School	$40,500	$224,000

Now that you understand your options to pay, what does it actually cost? Tuition is the big expense most people pay attention to. Many students start at community college the first two years to keep costs down. Also, if you go to school in your state, whether a two or four year institution, you may qualify for in-state tuition – a significantly lower rate than your out-of-state counterparts will pay. Your college will list in-state costs if they offer them.

If you need to move out of state, you may be able to achieve in-state status. This usually requires one year residency, a driver's license, not being claimed on your parents tax return, and working and filing taxes in the new state. There are usually minimums you have to earn with a cap on how much can be given to you by relatives. It may sound complicated, but I did it in Virginia, so it can be done. Your school website should list eligibility requirements for gaining in-state tuition.

Tuition is one thing, but there are other costs to consider:
1. On-campus housing costs with a meal plan
2. Rent, food, utilities if off-campus – assume 50% higher than on-campus costs if you can't find good information
3. Books and supplies – estimate $1000 per year
4. Phone plan
5. If needed, a car with insurance and gas
6. Spending money - $100/mo is not much for pizza, vices, clothes, etc.
7. One time costs – computer, equipment, desk

If it seems like a lot, it is. But, remember that you can work summers to help offset costs. You may also want to work while in school, but be warned that college will tax your time unlike anything you've ever seen in high school. It may be wise to go half-time if you plan to work while in school. Or, you may decide to take a semester off here and there to earn some cash (you have six months before you have to pay on your loans and returning to school resets the clock). Other options that may help are living at home while going to school or giving up your car for bus transportation. As mentioned, community college can also offset costs for the first two years.

If you recall, we started this discussion of costs with the question - can you afford school *for the long haul*? Let's say you manage to graduate with a Bachelor's degree in your chosen field. What are your expenses? How much do you have to pay in school loans? What will you earn to offset these costs?

Just as in college, you will need money for rent, food, and utilities as well as money for the phone, car, and spending. If you know where you will live in the U.S., you can use http://www.bestplaces.net/ to estimate cost of living. As a rule of thumb, the average rent for a one bedroom apartment in the U.S. is ~$800/month. Utilities ($300), food ($500), phone ($100), car and insurance ($300), and spending money ($200) easily gets you to about $2,200/month in expenses. (You can reduce some of this with a roommate.)

As for school loans, consider the worst case example. Let's say you were a "dependent" student that took 5 years to graduate and borrowed the full amount of $31,000 without making any payments thus far. With a 5% interest rate, you now owe just over $36,000. The loans have a 10 year repayment term once you are out of school, so you can expect to pay ~$450/month (a total of $54,000 when you are done).

If you were an independent student that borrowed $57,500 over 5 years, and let's assume the unlikely case that none of the loans were subsidized, you now owe ~$67,000 with a loan payment of ~$800/month (you'll pay $96,000 in the end).

You can save significant money in the long run by paying more than the minimum each month, but we'll start with the numbers given. You need to take home $2,650-$3,000 per month in pay, depending on your circumstances.

So, how much can you expect to take home? Well, you'll be starting out, so expect to be toward the lower end of the income bracket of your chosen field. Take 1/3 of that out for taxes and benefits. Benefits include low cost medical plans and contributions to your 401k should you choose to do so.

If we work backwards, a person needing to clear $2650/mo should aim to make $48,000 per year. If you need $3000/mo, then aim for $55,000 per year. You can quickly see why an unmarketable degree is bad news.

Using the examples above, you can start to estimate what tuition costs seem reasonable for your chosen career. These are not hard and fast numbers. Your living expenses after college don't have to be so high, and you don't have to borrow the full maximum amounts.

Finally, you may also consider if going to graduate school is a wise choice for you. You won't have to pay your loan for a bit longer, and your pay should be higher when you get out. This makes sense if you are in a major where you can go to school for free and/or with a small stipend to help with your cost of living. Otherwise, weigh the anticipated salary with the additional accumulated debt.

So, yes, college is expensive in addition to being hard. But, there are ways to pay for it. The military options can get you through debt-free to a paying job. The loan route is doable as long as you plan carefully. But, skipping college or some other training program altogether is the harder path in the long term. If you want to clear $2200/month, you need to find something that pays $40,000 per year (~$20/hour) and includes benefits. That is hard to do with only a high school degree.

It is my sincere wish that this information helps you make informed choices and solidify a plan. I wish you great success!

Other Books in This Series

The NOW 2 kNOW™ Mathematics Philosophy

If you are struggling with math, you don't have time to read a 400 page "help" book. You need the main concepts laid out plainly, simply, yet thoroughly.

The NOW 2 kNOW™ math texts give complete instruction in *80 pages or less*! In addition, you'll find *over 200* problems with *worked out* solutions!

See why math doesn't have to be hard or time-consuming to learn!

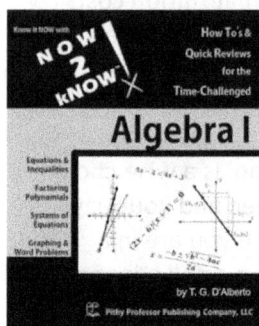

NOW 2 kNOW™ Algebra I

Algebra I is a course that requires a whole new way of thinking. Parents and students alike will love how the NOW 2 kNOW™ Algebra I text simplifies the subject while giving thorough and concise instruction.

NOW 2 kNOW™ Geometry
Geometry is the study of shapes and their parts. It is heavy in definitions, theorems, and proofs. The NOW 2 kNOW™ Geometry text organizes the subject's vast amount of information and provides numerous examples and practice with proofs!

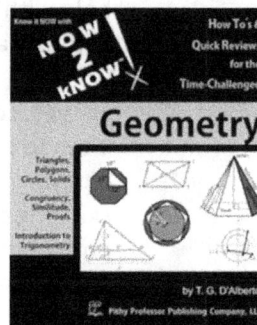

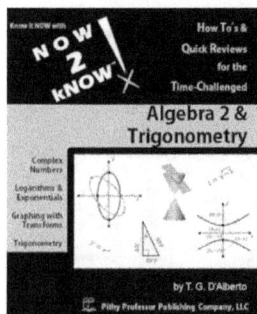

NOW 2 kNOW™ Algebra 2 & Trigonometry
Expand on the concepts of Algebra I and Geometry with this two course text! From imaginary numbers to logs and exponents to graphing and transformations to hyperbolic trigonometric functions, this text will have you on your way to Calculus in no time!

Don't Let a Fear of Math Stop You!